The Orphan Tsunami of 1700

みなしご元禄津波

A merchant's notebook, in an entry for January 1700, tells of a tsunami that lacked an associated earthquake in Japan. The shaking occurred instead along the northwest coast of North America. A French map compiled in 1720 shows what Europeans then knew of those shores.

jishin nite mo
earthquake

tsukamatsurazu
did not occur

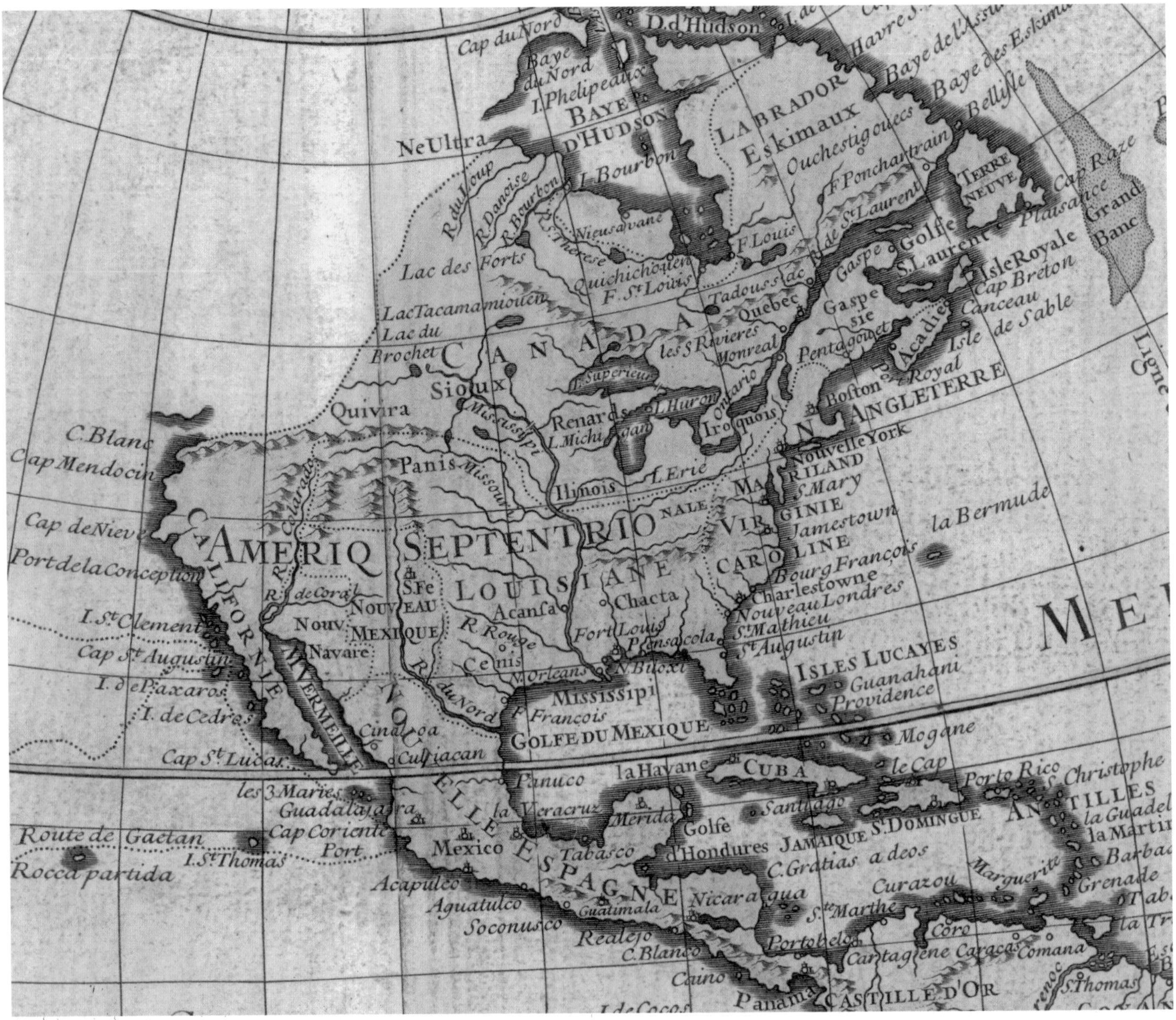

Quote from Moriai-ke "Nikki kakitome chō" (p. 50-52). Map by Guillaume Del'Isle, from University of Washington Libraries, Special Collections Division (p. 2, 5).

The Orphan Tsunami of 1700

みなしご元禄津波

Japanese clues to a parent earthquake in North America

親地震は北米西海岸にいた

Brian F. ATWATER   ブライアン・F・アトウォーター

MUSUMI-ROKKAKU Satoko   六角 聰子

SATAKE Kenji   佐竹 健治

TSUJI Yoshinobu   都司 嘉宣

UEDA Kazue   上田 和枝

David K. YAMAGUCHI   デイビッド・K・ヤマグチ

United States Geological Survey
Reston, Virginia

in association with

University of Washington Press
Seattle and London

Prepared in cooperation with the Geological Survey of Japan
(National Institute of Advanced Industrial Science and Technology),
the University of Tokyo, and the University of Washington

Published in 2005 simultaneously by the
U. S. Geological Survey (as Professional Paper 1707)
and University of Washington Press

Printed in Canada for University of Washington Press
12 11 10 09 08 07 06 05   5 4 3 2 1

University of Washington Press
P.O. Box 50096, Seattle, WA 98145
www.washington.edu/uwpress

An electronic version of the book, and of any updates to it, can
be found at http://pubs.usgs.gov/pp/pp1707/. For additional
USGS publications, visit http://www.usgs.gov/pubprod/.

The paper used in this book is acid-free and meets the
minimum requirements of American National Standard for
Information Sciences Permanence of Paper for Printed Library
Materials, ANSI Z39.48-1984.

Library of Congress Cataloging-in-Publication Data
The orphan tsunami of 1700 : Japanese clues to a parent
   earthquake in North America / Brian F. Atwater ... [et al.] =
   [Minashigo Genroku tsunami : oya-jishin wa Hokubei nishi
   kaigan ni ita / Buraian F. Atowota ... et al.].
         p. cm. -- (Professional paper ; 1707)
   Captions and table of contents also in Japanese.
   Parallel title and statement of responsibility also in
      Japanese characters.
   Includes bibliographical references and index.
   ISBN 0-295-98535-6 (acid-free paper)
   1. Paleoseismology--Northwest, Pacific.
2. Paleoseismology--Holocene. 3. Subduction zones--
Northwest, Pacific. 4. Tsunamis--Japan--History--18th
century--Sources. I. Title: Minashigo Genroku tsunami : oya-
jishin wa Hokubei nishi kaigan ni ita. II. Atwater, Brian F. III.
U.S. Geological Survey professional paper ; 1707.

QE539.2.P34 O77 2005
551.2'2'09795--dc22   2005050401

The authors receive no royalties from this publication.

LANGUAGE NOTES

みなしご　元禄　津波
minashigo　*Genroku*　*tsunami*

THE BOOK'S TITLE in Japanese, "Minashigo Genroku tsunami," means "The orphan tsunami of the Genroku era." In western calendars, the Genroku era began in 1688 and ended in 1704. Japanese written records tell of but one Genroku tsunami of remote origin. It dates to the year 1700 (p. 42).

親地震　は　北米　西海岸　に　いた
oya-jishin　*wa*　*Hokubei*　*nishi kaigan*　*ni*　*ita*

THE SUBTITLE, "Oya-jishin wa Hokubei nishi kaigan ni ita," means, "The parent earthquake was along the west coast of North America."

JAPANESE CITIZENS' NAMES appear in customary order, family name first. For clarity, the authors' family names contain small capital letters on the title page, pages 110-111, and the back cover.

TO WRITE JAPANESE WORDS in Roman letters we use a variant of the Hepburn system. The vowel sounds resemble those in Spanish: *a* resembles the first vowel in "mama," *e* the final vowel in "Santa Fé," *i* the second vowel in "police," *o* the first vowel in "José," and *u* the first vowel in "uno." The combination *ei* prolongs the *e* sound, as does *ii* for *i*. Prolonged *o* and *u* take macrons (*ō, ū*) except in internationalized words (Tokyo = Tōkyō). The *n* is pronounced *m* before *b* or *p* (Nambu, Sumpu) as it is in English (imbalance, empower). Additional changes in sound at the junctures between syllables or words are footnoted on pages 38, 52, 60, 68, and 78. A slight pause precedes a doubled consonant (*yokka*).

JAMES CURTIS HEPBURN (1815-1911), an American missionary, devised the system now employed widely, in modified form, to transcribe Japanese sounds into Roman letters. The standard dictionary by Nelson and Haig (1997) uses the Hepburn system. We hyphenate most counters (as in *niji-kken*, p. 39) but follow Nelson and Haig in closing compounds for the day counter *ka* (*yōka, yokka*).

Contents 目次

This book tells the scientific detective story of a giant earthquake and its trans-Pacific tsunami.

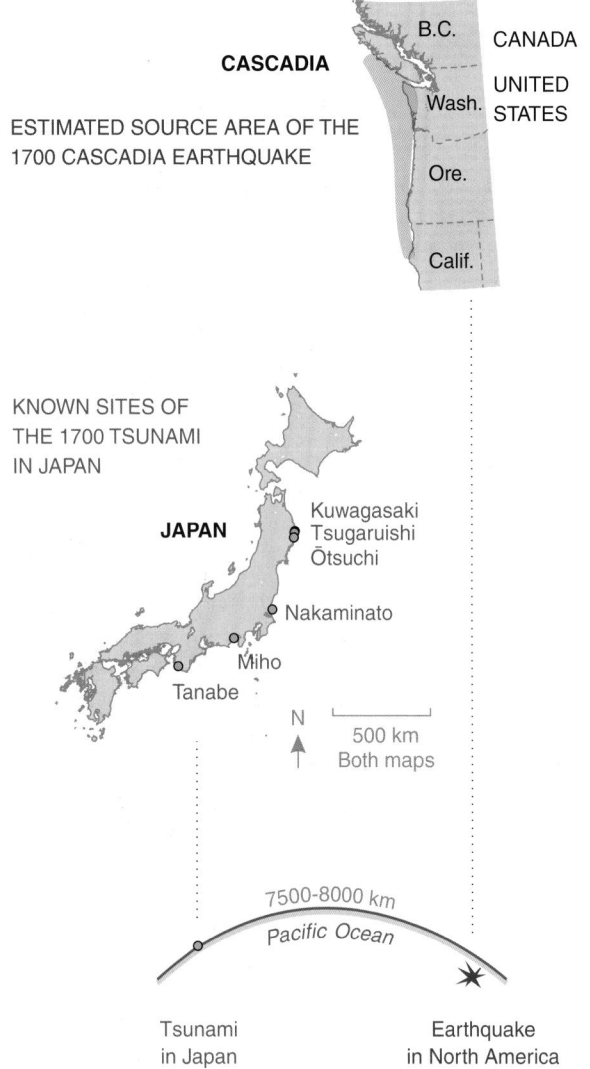

Part 1 illustrates geologic signs of enormous earthquakes and tsunamis at Cascadia, along the Pacific coast of North America from British Columbia to California.

Part 2 presents old Japanese writings about a tsunami of mysterious origin that caused flooding and damage in January 1700 from Kuwagasaki in the north to Tanabe in the south.

Part 3 links this orphan tsunami to a Cascadia earthquake and to seismic hazards in the western United States and Canada.

The Orphan Tsunami of 1700

みなしご元禄津波

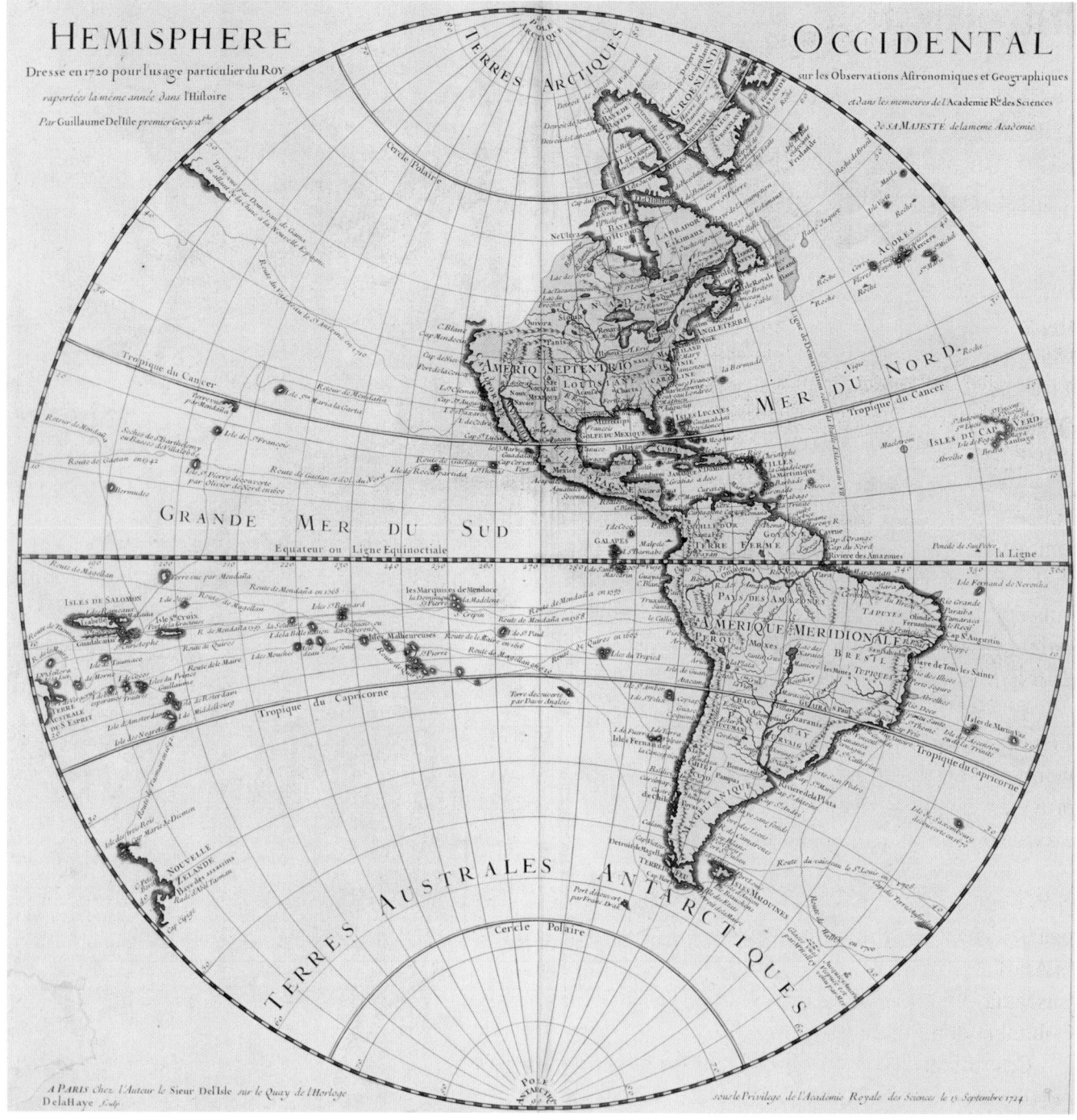

Still-uncharted shores in North America were home to a giant earthquake and its tsunami in the year 1700, two decades before France's royal geographer compiled this map. In North America, the catastrophe left traces on the landscape and probably in the oral histories of native people. Across the Pacific, the tsunami entered Japan's written history as a sea flood without local cause. Three centuries later, the combined clues would reveal that the North American earthquake probably attained magnitude 9.

Introduction はじめに

OUTSIDERS SCARCELY KNEW of northwestern North America in the year 1700. Leading European geographers of the time left that part of the map blank. Not until 1741 would Russians land in Alaska. From there to Oregon's Cape Blanco, the coast would remain uncharted until Spanish and English expeditions of the 1770s.

Across the Pacific Ocean in Japan, unusual seas ran ashore in 1700. People wrote of the effects: flooded fields, wrecked houses, a fire, a shipwreck, evacuation, fright. Having felt no earthquake beforehand, some writers called the flooding a "high tide" and most resisted calling it a tsunami. None could have known that a seismic shift on a North American fault had set off a train of trans-Pacific waves. Far from its parent earthquake, the tsunami of 1700 was an orphan.

The 1700 tsunami in Japan would remain an orphan for nearly three hundred years. The North American fault at its source would go unnoticed until the last decades of the 20th century. Today the fault is charted, and an earthquake on it is regarded as the orphan's parent. This kinship gives the earthquake an exact date (January 26, 1700) and an estimated size (magnitude 8.7-9.2) that spur precautions against future earthquakes and tsunamis in the United States and Canada.

THE INDIAN OCEAN TSUNAMI of December 26, 2004, reminded the world of what an earthquake of magnitude 9 can do. Earth rarely provides such reminders; only three 20th-century earthquakes reached or exceeded magnitude 9.0 worldwide.

The Indian Ocean disaster, by affecting areas from southeast Asia to Africa, raised concern about earthquake and tsunami hazards around the planet. The disaster reminded North Americans of such hazards not only in Alaska, struck in 1964 by an earthquake of magnitude 9.2, but also at Cascadia—the region west of the Cascade Range from southern British Columbia to northern California.

Cascadia is home to a gently inclined boundary between two of the moving tectonic plates that make up Earth's outer shell. The shallow, mostly offshore part of the boundary is the fault that ruptured in 1700. What losses will Cascadia sustain the next time it breaks? A scenario printed in 2005, several months after the Indian Ocean disaster, gives an idea of what to expect.

The scenario begins with an earthquake of magnitude 9.0. Strong shaking lasts for minutes along the Pacific coast in British Columbia, Washington, Oregon, and California. The main coastal highway, U.S. 101, becomes largely impassable, and landslides "sever highway travel between the coast and inland areas." Thus isolated, coastal residents "have to do much of the work of rescuing those trapped in the rubble."

The expected damage extends inland to Vancouver, Seattle, and Portland. In this urban corridor, "utilities and transportation lines in some areas could be disrupted, perhaps for months." Damage to tall buildings "could lead to significant fatalities in downtown areas."

These risks used to be unthinkable. Cascadia has no written records of homemade earthquakes larger than magnitude 7.5, nor of trans-oceanic tsunamis generated in its backyard. However, the region does have geologic records of great earthquakes—shocks of magnitude 8 or larger—and of tsunamis they spawned. It is the most recent of these Cascadia tsunamis that entered written history in Japan.

RECOGNIZING A HAZARD is just the first step toward dealing with it. Next, the hazard must be defined well enough for practical precautions to be devised and put into effect.

Discoveries about the orphan tsunami of 1700 helped drive this process at Cascadia. Earth science in North America revealed earthquake and tsunami hazards that Japanese history sharply defined. The findings spurred precautionary steps like the mapping of areas that future Cascadia tsunamis may flood and the posting of evacuation signs. The safeguards also include teaching schoolchildren the basics of tsunami survival: If you feel a strong earthquake, run to high ground. If the sea recedes strangely, run to high ground. If a tsunami ensues, stay on high ground; its first wave probably won't be the last—or the highest.

If only such precautions could have been taken around the Indian Ocean before its 2004 disaster. Most of the victims experienced the earthquake, which was felt even in Thailand and Sri Lanka. Many saw the sea withdraw before the first damaging wave. Some thought the first wave would be the last. Almost everyone was surprised by the earthquake's magnitude and by the tsunami's height and reach. The 2004 earthquake and tsunami were outsize events with hardly any known precedent in the Indian Ocean's past.

IN THIS BOOK we use the past to help warn of outsize earthquakes and tsunamis of the future. We assemble clues from both sides of the Pacific to establish precedent for a giant Cascadia earthquake and its tsunami. We tell the detective story behind some of the recent precautions against earthquakes and tsunamis in western North America.

Five of us were among the detectives. Ueda and Tsuji identified, verified, and correlated several of the Japanese accounts of an orphan tsunami from 1700. Satake recognized this tsunami's probable link to North American geology and estimated the parent earthquake's size. Atwater discovered some of that geology and Yamaguchi led in dating it, with tree rings, to a 10-month window that contains the orphan tsunami's time.

The discoveries thrilled and astonished us—and they still do. But they also bring to mind the Indian Ocean disaster. How many actual orphans did the tsunami of 1700 create?

SIGNS OF CATASTROPHE in 1700 can still be seen in sediments and trees of northwestern North America and in archives of shogunal Japan. Having been privileged to examine these clues, we try to tell the story through them.

The Japanese archives tell of the 1700 tsunami in the words of magistrates, merchants, and peasants. We reproduce each account in full and, guided by linguist Musumi-Rokkaku, state its literal meaning in English. We also explore how each account came to be written and preserved, and how earthquake historians learned of it. Today's North American precautions against earthquakes and tsunamis are founded, in part, on these minutiae of Japanese history.

THE MAP on the frontispiece and page 2, "Hemisphere occidental," was compiled in 1720 and published in 1724 by Guillaume Del'Isle, then France's foremost cartographer (Portinaro and Knirsch, 1987, p. 314; French, 1999, p. 353-354). Del'Isle began publishing maps in 1700, gained a reputation for accuracy, and was appointed royal mapmaker—Premier Géographe du Roi—in 1718. University of Washington Libraries, Special Collections, UW23622z.

EARLIER MAPS that leave northwestern North America blank include "Nova totius terrarum orbis tabula" by Frederick de Wit, 1665; "Novissima totius terrarum orbis tabula" by John Seller, ca. 1673; de Wit's "Totius Americae descriptio," 1690, and "A new map of America" by Edgar Wells, 1700 (Portinaro and Knirsch, 1987, p. 186-209). Hayes (1999) chronicles the European discovery of northwestern North America by presenting the explorers' maps.

FAMILIAR WESTERNERS OF 1700 include Antonio Vivaldi, Johann Sebastian Bach, George Frideric Handel; Daniel Defoe, Jonathan Swift, Alexander Pope; Issac Newton, Gottfried Wilhelm von Leibniz, Jakob Bernoulli, Edmond Halley, Gabriel Daniel Farenheit; John Locke, Voltaire, Montesquieu; Rob Roy, William "Captain" Kidd; and Peter the Great. Charles Perrault's "Little Red Riding Hood" appeared in 1697. In 1700 London was Europe's largest city with a population of 550,000. In England's North American colonies, residents of Boston numbered 7,000, and New York was a town of 5,000. The school later renamed Yale University opened in 1702. Benjamin Franklin was born six years after the 1700 earthquake. Sources: Pascoe (1991), Garruth (1993), and Williams (1999).

"A MAGNITUDE 9.0 EARTHQUAKE SCENARIO" was prepared by a panel of scientists, engineers, and officials from government and industry, the Cascadia Region Earthquake Workgroup (2005). The scenario does not include numerical estimates of losses of life or property.

THE 2004 SUMATRA-ANDAMAN EARTHQUAKE was felt, at low intensity, in Sri Lanka, peninsular India, Myanmar, Malaysia, and Thailand. Estimates of the earthquake's moment magnitude range from 9.0 (for seismic waves of 300-second period) to 9.3 (including waves of periods >500 seconds). By the criteria used to estimate the size of the 20th century's largest earthquakes (graph, p. 98), the 2004 Sumatra-Andaman earthquake attained magnitude 9.0 (Lay and others, 2005).

THE INDIAN OCEAN has a written history of dozens of tsunamis since the middle of the 18th century (http://www.ngdc.noaa.gov/spotlight/tsunami/tsunami.html). One of the largest of these was generated in 1833 during an earthquake of estimated magnitude 8.8-9.2 along the west coast of Sumatra (Zachariasen and others, 1999). Its rupture area lies a few hundred kilometers south of the southern end of the 2004 break. Northern parts of the 2004 rupture are the most likely sources of earthquakes of magnitude 8 in 1847, 1881, and 1941 (p. 101). A tsunami in the Bay of Bengal is known to have accompanied the earthquake of 1881 (Bilham and others, 2005, p. 304).

A TSUNAMI-SURVIVAL GUIDE by Atwater and others (1999) mentions Pacific and Atlantic hazards but not the Indian Ocean.

DOCUMENTARIES on findings central to this book include:
"The quake hunters"
    http://www.films.com/id/10444
"Cascadia, the hidden fire"
    http://www.globalnetproductions.com/products.html
"The next megaquake"
    http://www.bbc.co.uk/sn/tvradio/programmes/horizon/megaquake_qa.shtml
"Unearthing proof of a tsunami in the Pacific Northwest"
    http://www.npr.org/templates/story/story.php?storyid=4629401

Spruce cannot live where barnacles and rockweed cling. Trees like these, killed by tidal submergence, fueled discoveries in the late 1980s and early 1990s about earthquake hazards at Cascadia.

Willapa Bay at low tide, 4 km north of Oysterville, Washington, 1990. The striped handle is 0.5 m long. A man digs behind stumps at upper right.

Part 1
Unearthed earthquakes　発掘された地震痕跡

THROUGH MOST OF THE 20TH CENTURY, North America's Cascadia region was thought incapable of generating earthquakes larger than magnitude 7.5. Any tsunami striking the region's coasts would come from afar, leaving hours for warning and evacuation. Yet by century's end, Cascadia had its own recognized source of earthquakes of magnitude 8 to 9 and of tsunamis that would reach its shores in a few tens of minutes.

That recognition began in the early 1980s. Earth scientists were then beginning to debate Cascadia's potential for great earthquakes—shocks of magnitude 8 or higher. Despite hints from oral histories of native peoples, there seemed no way to learn whether great earthquakes had ever struck the region.

Fortunately, the earthquakes had written their own history. They wrote it most clearly in the ways that great earthquakes of the 1960s in Chile and Alaska wrote theirs—by dropping coasts a meter or two, by sending sand-laden sea water surging across the freshly lowered land, and by causing shaken land to crack.

Those geologic records soon gave Cascadia a recognized history of great earthquakes. In the late 1980s, at bays and river mouths along Cascadia's Pacific coast, researchers found the buried remains of marshes and forests that subsidence had changed into tidal mudflats. They also found that the burial began with sand delivered by tsunami or erupted in response to shaking. In a few places they even found the hearths of native people who had used the land before its submergence and burial.

But researchers quickly reached an impasse in this attempt to define, from events recorded geologically, Cascadia's earthquake and tsunami hazards. How great an earthquake should a school or hospital be designed to withstand? How large a tsunami should govern evacuation plans on the coast? There seemed no way to know whether Cascadia's plate-boundary fault can unzip all at once, in a giant earthquake of magnitude 9, or whether it must break piecemeal, in series of lesser shocks.

Earthquake potential 地震の可能性

Can Cascadia do what other subduction zones have done?

CASCADIA'S CONVERGING PLATES pose a triple seismic threat. The subducted Juan de Fuca Plate contains sources of earthquakes as large as magnitude 7. Large earthquakes can also radiate from faults in the overriding North America Plate. And the enormous fault that forms the boundary between the plates can produce great earthquakes, of magnitude 8 or 9.

This current picture began taking form in the 1960s, when early ideas about continental drift and seafloor spreading came together as the theory of plate tectonics. The Juan de Fuca Plate was identified as a remnant of a larger tectonic plate that had mostly disappeared beneath North America during 150 million years of subduction.

By the early 1980s, geophysicists had shown that the Juan de Fuca Plate continues to subduct at an average rate of 4 meters per century. But there was no consensus on how the plates move past one another. The plate boundary lacked a recognized history of earthquakes, even at the shallow depths where the rocks might be cool and brittle enough to break (pink in block diagram and map, right).

An earthquake in 1985 provided disturbing images of what can happen when such a plate boundary fails. On September 19th of that year, a subduction earthquake of magnitude 8 generated seismic waves that devastated Mexico City, 400 km from the earthquake source (facing page, top). More than 300 modern buildings collapsed or were damaged beyond repair, 10,000 lives were lost, and another 300,000 persons were left homeless. Could a great Cascadia earthquake have similar effects at inland cities like Vancouver, Seattle, and Portland?

Though few Earth scientists were then taking the idea seriously, some broached the possibility of a Cascadia earthquake of magnitude 9. Cascadia looked like it might have as much source area as the 1964 Alaska earthquake, of magnitude 9.2 (compare the Cascadia and Alaska maps on these two pages). It was even possible to imagine a Cascadia earthquake as large as the 1960 Chile mainshock, the 20th century's largest earthquake at magnitude 9.5.

THE THEORY OF PLATE TECTONICS holds that Earth's outer shell consists of moving plates composed of crust and rigid upper mantle (Sullivan, 1991; Oreskes, 2003). Riddihough (1984) used seafloor magnetic anomalies, first mapped by Raff and Mason (1961), to reconstruct the past 7 million years of convergence between the Juan de Fuca and North America Plates. Tanya Atwater's animations of these and other plate motions can be downloaded at http://emvc.geol.ucsb.edu/.

EARLY IDEAS on Cascadia's great-earthquake potential were reviewed by Heaton and Hartzell (1987), Rogers (1988), Hyndman (1995), and Rogers and others (1996). Heaton and Hartzell (1986, p. 688-694) proposed that Cascadia might produce earthquakes as large as the 1964 Alaska and 1960 Chile events. Beck and Hall (1986) inferred that long-lasting seismic waves, and their resonance in ancient lake deposits, contributed to Mexico City's earthquake losses in 1985.

EARTHQUAKE SOURCES AT CASCADIA

Inland cities

Cascade volcanoes

Pacific Ocean

1700

c. 900

2001

North America Plate

Juan de Fuca Plate

Year of most recent major quake

Potential source of great earthquakes

Plate boundary

North America Plate

Juan de Fuca Plate

100 km

Mantle Crust

130°W   120°

B.C.

50°N   Vancouver

WA
Seattle

Juan de Fuca Plate

Portland
OR

North America Plate

Pacific Plate

CA

40°   Pacific Ocean

San Andreas Fault

0   500 km   North at 125°W

Potential source of great earthquakes at plate boundary (p. 99).

Seaward edge of subduction zone Seafloor projection of gently inclined fault between subducting and overriding plates

Spreading ridge Submarine mountain range where injected magma forms new oceanic crust

Vertical fault

EARTHQUAKE ANALOGS ON FACING PAGE

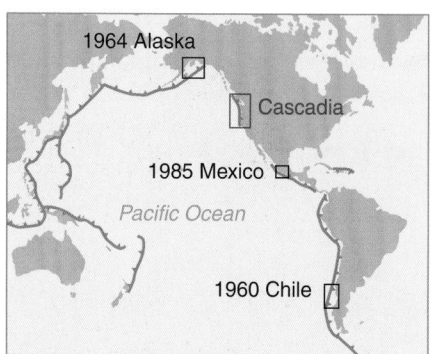

1964 Alaska

Cascadia

1985 Mexico

Pacific Ocean

1960 Chile

Seaward edge of subduction zone

MEXICO, 1985, M 8.1

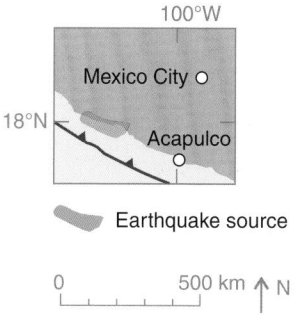

Earthquake source

0 — 500 km ↑N

Tall buildings shortened by collapse of entire stories, Mexico City.

ALASKA, 1964, M 9.2

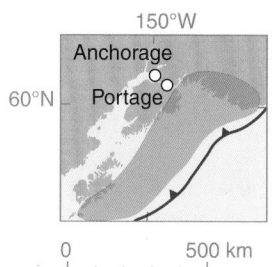

0 — 500 km

The maps on this page are at the same scale as the Cascadia map on the facing page.

Remains of highway bridge, Portage. Wooden pilings pierce deck they used to support. Aerial view, p. 14.

School torn by landslide, Anchorage.

CHILE, 1960, M 9.5

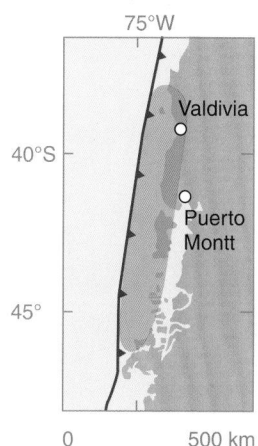

0 — 500 km

Cracked street and variably damaged buildings, Valdivia.

Broken sewer main, Puerto Montt.

EARTHQUAKE SOURCES (fault rupture areas) inferred from aftershocks for Mexico (UNAM Seismology Group, 1986) and Alaska (Plafker, 1969, p. 6) and from aftershocks and land-level changes for Chile (Cifuentes, 1989, p. 676).

PHOTOS from the Karl V. Steinbrugge collection, National Information Service for Earthquake Engineering, University of California, Berkeley. Photographers: Karl Steinbrugge, Rodolfo Schild (Valdivia), and anonymous (Portage).

Tsunami potential 津波の可能性
Is Cascadia further threatened by its own Pacific tsunamis?

BEFORE

THE POTENTIAL for a great Cascadia earthquake carries with it the threat of an ensuing tsunami. And the tsunami that follows a great subduction earthquake often does more harm than the earthquake itself.

The giant Chilean earthquake of 1960, for instance, left many houses standing above the earthquake source (Valdivia photo, previous page). But the tsunami that followed erased entire villages, including Queule (above). In total, the 1960

tsunami took an estimated 1,000 lives in Chile. It also claimed 61 in Hawaii and 138 in Japan.

You can make a tsunami in a bathtub by sweeping your hand through the water. During a great subduction earthquake, the role of the hand is played by a moving tectonic plate. The plate displaces water from beneath by warping the sea floor (below). It is this tectonic warping, not the seismic shaking, that acts as the hand in the tub.

MAKING A TSUNAMI

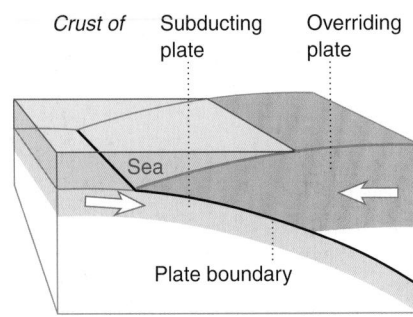

OVERALL, a tectonic plate descends, or "subducts," beneath an adjoining plate. But it does so in a stick-slip fashion.

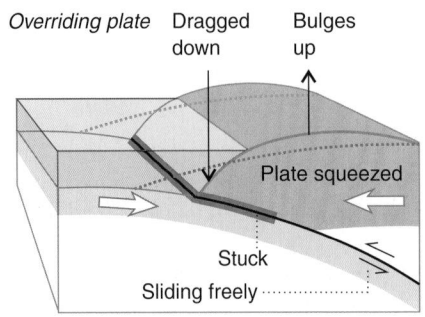

BETWEEN EARTHQUAKES the plates slide freely at great depth, where hot and ductile. But at shallow depth, where cool and brittle, they stick together. Slowly squeezed, the overriding plate thickens.

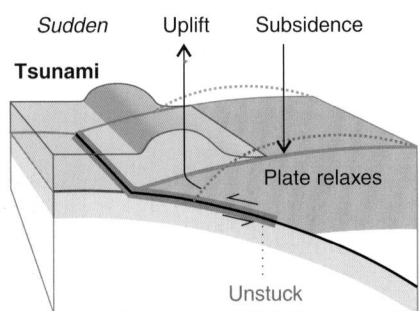

DURING AN EARTHQUAKE the leading edge of the overriding plate breaks free, springing seaward and upward. Behind, the plate stretches; its surface falls. The vertical displacements set off a tsunami.

AFTER

Such sea-floor deformation during the 1964 Alaska earthquake generated a tsunami that reached Cascadia's shores four hours later. Warnings had been issued, and the tsunami had lost height by spreading out as it radiated toward far reaches of the Pacific Rim. Even so, the Alaskan waves swept four children off an Oregon beach and killed another dozen persons in northern California.

Before and after the 1960 tsunami in Queule, Chile, 2 km inland from the sea. Warned by the M 9.5 earthquake or by early signs of the tsunami, most residents managed to reach high ground before surges of seawater swept away their homes.

A QUEULE SURVIVOR recalled seeing the village's houses floating off like an armada (Atwater and others, 1999, p. 5, 15). Photographs by Wolfgang Weischet (1963, p. 1245-1246); prints courtesy of Pierre Saint-Amand.

TEXTBOOKS on tsunamis include Murty (1977), Dudley and Lee (1998), and Bryant (2001). On historical tsunamis in the United States, see Lander and others (1993) for the Pacific coast and Lockbridge and others (2002) for the Atlantic.

Downwarped coast

THE PLATE MOTION that drives a tsunami can also lower a coast (cartoons, opposite). When fault slip during the 1960 earthquake stretched the overriding South America plate, the Earth's surface fell throughout a belt 1,000 km long (map, right). In Queule, near the axis of the downwarp, the entire landscape dropped 2 meters. Hence tides cover former riverbanks in the postearthquake photo above.

Similar coastal subsidence accompanied the 1964 Alaska earthquake (p. 14) and several earthquakes in southwest Japan (p. 91). Coastal subsidence accordingly provides pivotal evidence for the past occurrence of great Cascadia earthquakes—clues Earth scientists began using in the late 1980s (p. 16).

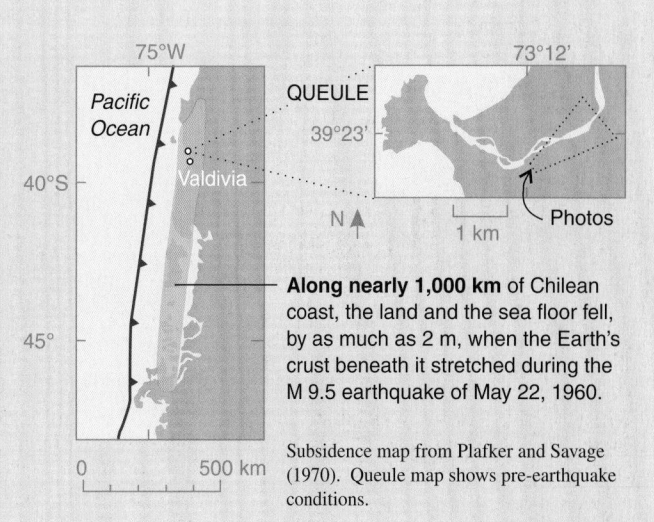

Along nearly 1,000 km of Chilean coast, the land and the sea floor fell, by as much as 2 m, when the Earth's crust beneath it stretched during the M 9.5 earthquake of May 22, 1960.

Subsidence map from Plafker and Savage (1970). Queule map shows pre-earthquake conditions.

Flood stories 洪水の言伝え

Cascadia's own tsunamis may have entered Native American lore.

OLD WRITINGS FROM CASCADIA offer few hints that the region's subduction zone produces great earthquakes or tsunamis. Such events are unknown from the records of early explorers like Bruno de Hezeta y Dudagoifia, who mapped the mouth of Washington's Quinault River in 1775; James Cook, who named Cape Flattery a few years later; and George Vancouver, who surveyed Puget Sound, Grays Harbor, and the lower Columbia River in the early 1790s. The Lewis and Clark Expedition recorded no signs of Pacific coast earthquakes or tsunamis while exploring that river in 1805 and 1806.

The oral traditions of Cascadia's native peoples, however, tell of flooding from the sea. The example at right, one of the first written, comes from James Swan's diary for a rainy Tuesday in January 1864 at Neah Bay, Washington Territory, home of the Makah tribe. Swan's informant, Billy Balch, was a Makah leader.

Balch recounts a sea flood in the "not very remote" past. It began by submerging the lowland between Neah Bay and the Pacific Ocean. Next, the water receded for four days. Rising again "without any swell" the sea covered all but the highest ground on both sides of the Strait of Juan de Fuca. It dispersed tribes, stranded canoes in trees, and caused "numerous" deaths. "The same thing happened" at Quileute, 50 km south of Neah Bay.

Balch mentions no earthquake. Did the sea flood have a remote origin, like the tsunami from Alaska in 1964? Or did a tsunami of nearby origin prove more memorable than the Cascadia earthquake that triggered it?

DIARY OF JAMES SWAN FOR JANUARY 12, 1864 ▷

Billy also related an interesting tradition. He says that "ankarty" but not "hias ankarty" that is at not a very remote period the water flowed from Neeah bay through the Waatch prairie, and Cape Flattery was an Island. That
5 the water receded and left Neeah Bay dry for four days and became very warm. It then rose again without any swell or waves and submerged the whole of the cape and in fact the whole country except the mountains back of Clyoquote. As the water rose those who had canoes put their
10 effects into them and floated off with the current which set strong to the north. Some drifted one way and some another and when the waters again resumed their accustomed level a portion of the tribe found themselves beyond Nootka where their descendants now reside and are known by the
15 same name as the Makahs—or Quinaitchechat. Many canoes came down in the trees and were destroyed, and numerous lives were lost. The same thing happened at Quillehuyte and a portion of that tribe went off either in canoes or by land and formed the Chimakum tribe
20 at Port Townsend.

There is no doubt in my mind of the truth of this tradition. The Waatch prairie shows conclusively that the waters of the ocean once flowed through it. And as this whole country shows marked evidence of volcanic influences there is
25 every reason to believe that there was a gradual depression and subsquent upheaval of the earths crust which made the waters to rise and recede as the Indian stated.

The tradition respecting the Chimakums and Quillehuytes I have often heard before from both those tribes.

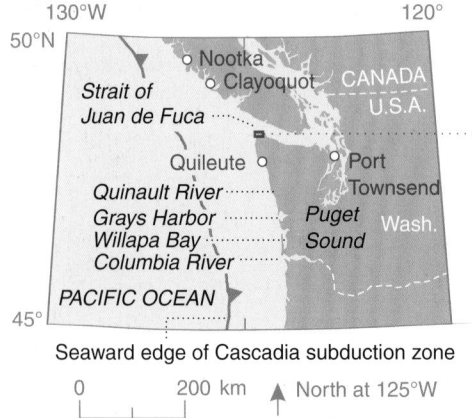

The diary refers to the places in blue.

EARLY EXPLORATION summarized from Hayes (1999, p. 37-93).

ORAL TRADITIONS of Cascadia earthquakes and tsunamis are summarized in a comprehensive collection by Ludwin and others (2005). Swan (1870, p. 57-58) recounts Billy Balch's story; Doig (1980, p. 62-65) puts it in context of Swan's diary-keeping. Swan (1857, p. 417) defines the Chinook jargon *hias* (or *hyas*).

MAP at right excerpted from "Makah Indian Reservation in Washington Territory by J.G. Swan, 1862" (National Archives and Records Administration, RG 75, #995). Shorelines probably based on mapping by the U.S. Coast Survey.

BILLY BALCH'S family history is recounted by Goodman and Swan (2003), who give his Makah name as Yelakub.

JAMES G. SWAN (1818-1900) lived among Indians of Shoalwater (now Willapa) Bay in the early 1850s and among the Makah in the 1860s (McDonald, 1972). He wrote newspaper articles and books about these people and their land (Swan, 1857, 1870, 1971). He also penned two and a half million words in diaries that span forty years (Doig, 1980). The excerpt is from University of Washington Libraries, Special Collections, UW19484z and UW19485z.

Billy also related an interesting tradition. He says that "ankaty" but not "hias ankaty" that is at not a very remote period the water flowed from neah bay through the Waatch prairie, and Cape Flattery was an Island. That the water receded and left Neah Bay dry for four days and became very warm. it then rose again without any swell or waves and submerged the whole of the cape and in fact the whole country except the mountains back of Clyoquot. As the water rose those who had canoes put their effects into them and floated off with the current which set strong to the north. Some drifted one way and some another and when the waters again resumed their accustomed level a portion of the tribe found themselves beyond Nootka where their descendants now reside and are known by the same name as the makahs or Quenaitchechat.

Many canoes came down in the trees and were destroyed and numerous lives were lost. The same thing happened at Quillehuyte and a portion of that tribe went off either in canoes or by land and formed the Chemakum tribe at port Townsend

There is no doubt in my mind of the truth of this tradition. The Waatch prairie shews conclusively that the waters of the ocean once flowed through it. and as this whole country shews marked evidence of volcanic influences there is every reason to believe that there was a gradual depression and subsequent upheaval of the earth crust which made the water to rise and recede as the Indians stated.

The tradition respecting the Chemakums and Quillehuyts I have often heard before from both these tribes

Alaskan analog アラスカの例

Ghost forests and a buried soil naturally record the 1964 earthquake.

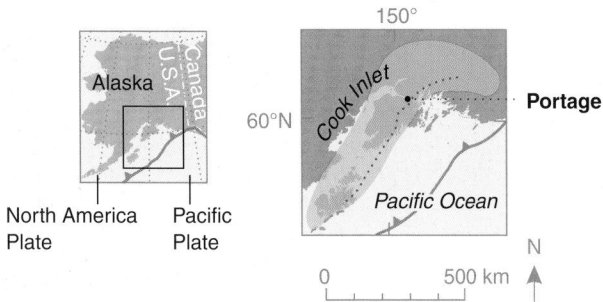

Alaska
Canada
U.S.A.
North America Plate
Pacific Plate
150°
60°N
Cook Inlet
Portage
Pacific Ocean
N
0
500 km

Tectonic subsidence during the 1964 Alaska earthquake Maximum 2.3 m; axis at dotted line. Inferred cause described, p. 10-11.

Seaward edge of subduction zone Low-angle fault at plate boundary; teeth point down the fault plane, as on p. 8.

SOME EARTHQUAKES write their own history. In a classic example, the giant Alaska earthquake of March 27, 1964, was accompanied by regional subsidence that lowered vegetated land into Cook Inlet. The results remain easy to see at Portage, near the axis of the earthquake's downwarp. Trees stand dead because the land subsided 1.5 meters during the earthquake—far enough to admit tides into former meadows, willow thickets, cottonwood groves, and stands of Sitka spruce. Tides brought in silt and sand that buried this former landscape in the first decade after the earthquake.

Like the Portage garage, opposite, the spruce victims were falling by 1998. Their stumps, however, remain in growth position, entombed by the tidal silt. These serve as natural archives of the 1964 earthquake—and as a guide to identifying signatures of past great earthquakes at Cascadia.

TIDAL FLOODING A FEW WEEKS AFTER EARTHQUAKE, TWENTYMILE RIVER

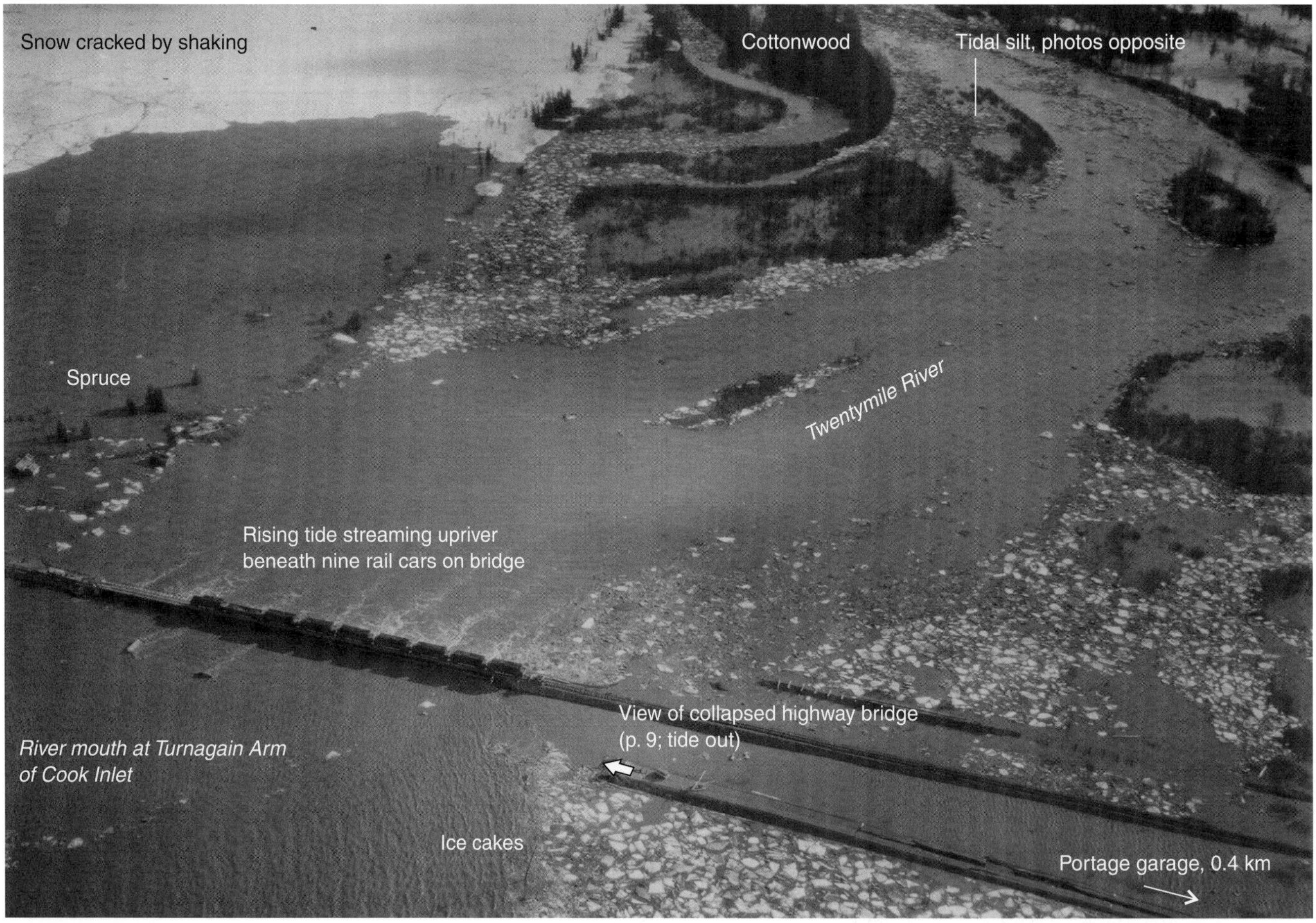

Snow cracked by shaking
Cottonwood
Tidal silt, photos opposite
Spruce
Twentymile River
Rising tide streaming upriver beneath nine rail cars on bridge
View of collapsed highway bridge (p. 9; tide out)
River mouth at Turnagain Arm of Cook Inlet
Ice cakes
Portage garage, 0.4 km

AIRPHOTO from U.S. Army, Mohawk series M-64-82. Probably taken during high tide of April 14, 1964 — 18 days after the March 27 earthquake.

THE TECTONIC COMPONENT of the subsidence near Portage amounted to 1.5-1.7 m (Plafker, 1969, plate 1). The railroad grade settled another 1 m, on average, in response to seismic shaking (McCulloch and Bonilla, 1970, p. 81).

THE SUBSIDENCE at Portage created intertidal space that the silt and sand filled (Ovenshine and others, 1976). Much of the fill dates from the first months after the earthquake, when individual high tides left layers as much as 2 cm thick (opposite; Atwater and others, 2001b). The deposition was speeded by a 10-m tide range and ample sources of sediment (Bartsch-Winkler, 1988).

GHOST FOREST AT PORTAGE GARAGE

Summer 1964

Spruce dying

PORTAGE GARAGE

High tide covers recently subsided land

1998

Few trunks still standing

Men

Shrubs cover land rebuilt by tidal silt (example below)

Before earthquake

Several months after earthquake

Several decades after earthquake

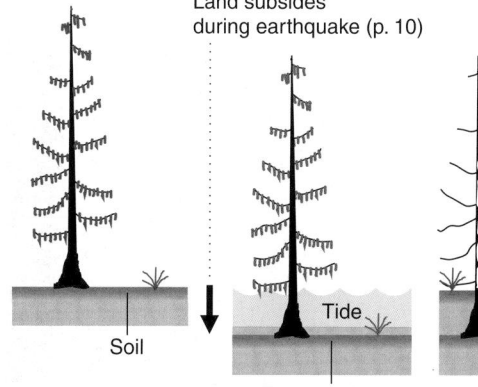

Land subsides during earthquake (p. 10)

Soil

Tide

Buried soil

Silt

By lowering land into a bay or river mouth, subsidence during an earthquake produces a lasting record of the earthquake's occurrence.

TIDAL SILT AND SAND ABOVE BURIED SOIL, TWENTYMILE RIVER

Ground surface in 1998

Tidal silt and sand, 1.5 m thick, mostly deposited in the first few years after the 1964 earthquake

Detail below

Ground surface before the earthquake

10 cm

Sand and silt of one high tide, probably April 1964

Buried soil

Sunken shores 海岸の沈降

Land occasionally drops along Cascadia's Pacific coast.

Copalis River; very high fair-weather tide, December 1997

GHOST FOREST IN TIDAL MARSH

IT WAS ONLY A MATTER OF TIME before someone would recognize Portage look-alikes at Cascadia. In the late 1980s, spurred by controversy about Cascadia's great-earthquake potential, geologists checked bays and river mouths along Cascadia's Pacific coast. At nearly every one they found evidence that land had dropped.

These signs of subsidence include ghost forests—groves of weather-beaten trunks that stand in tidal marshes of southern Washington. First documented in the early 1850s, they are composed entirely of western red cedar, a long-lived conifer known for rot-resistant wood.

More common are victim trees preserved only as stumps beneath the marshes. Thousands of such stumps can be seen in banks of tidal streams in southern Washington (example opposite), hundreds more at estuaries in Oregon and northern California. The main victim is Sitka spruce—the species whose rotting trunks were falling at Portage in the fourth decade after their deaths in 1964 (p. 15).

Most common of all are the buried remains of tidal marshes. In streambanks and sediment cores, muddy tidal deposits abruptly overlie peaty marsh soils.

THE LOWERING OF LAND by Cascadia earthquakes has been inferred in dozens of reports. Recent examples include details from Oregon (Kelsey and others, 2002; Witter and others, 2003; Nelson and others, 2004) and Washington (Atwater and others, 2004) and a regional compilation (Leonard and others, 2004).

IN THE UPPER PHOTO, spruce saplings live high on the tallest of the dead trees. Several of the tall trunks fell between 1997 and 2003.

JAMES GRAHAM COOPER (1830-1902), wintered at Shoalwater (now Willapa) Bay in 1853-1854, while serving as naturalist for a railway survey. He described the bay's ghost forests of western red cedar to illustrate the wood's durability. He inferred that the trees had spent their lives "above high-water level, groves of this and other species still flourishing down to the very edge of inundation" (Cooper, 1860, p. 26). As to what killed the trees, Cooper proposed gradual sinking into quicksand. Now it is clear that the land dropped suddenly (evidence opposite), and that subsidence resulted from stretching of solid rock (right cartoon, p. 10).

Ordinary tide, August 1991

SUBSIDED SITES ALONG CASCADIA'S PACIFIC COAST

○ **Bay or river mouth** where buried soils record earthquake-induced submergence in the past 5000 years.

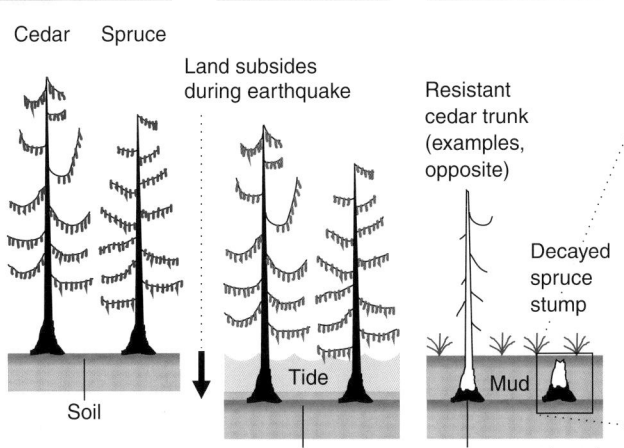

Before subsidence | First years after subsidence | Centuries later

Cedar Spruce

Land subsides during earthquake

Resistant cedar trunk (examples, opposite)

Decayed spruce stump

Tide

Soil

Buried soil   Bark preserved (p. 96)

Mud

SPRUCE STUMP EXHUMED IN BANK OF TIDAL STREAM

Modern salt marsh

Mud deposited by tides

0.5 m

Stump

Buried forest floor

Naselle River, Willapa Bay

Whodunit

WHAT ALLOWED TIDES to kill forests and bury marshes along Cascadia's Pacific coast? At first, in the late 1980s, geologists couldn't rule out gradual rise of the sea. But soon they convicted abrupt fall of the land that was accompanied by tsunami and shaking (p. 18 and 20), and which happened in the same decades at different places along a shared fault (p. 24).

Sudden subsidence provides a simple explanation for tree rings like those at right. The rings record the final decades of life for a Sitka spruce killed by postearthquake submergence at Willapa Bay, like the one in the photo above. Wide to the end, the rings suggest that the tree was healthy right up to the time of its death. The rings show no sign of lengthy suffering from gradual drowning and salt-water poisoning from a drawn-out sea-level rise.

Sudden lowering of land also explains remarkable preservation of buried marsh soils at Cascadia. Some of the soils retain delicate remains of plants that had been living on them at the time of submergence. Tidal-flat mud above such soils entombed herbaceous leaves and stems, in growth position, before they had time to rot. Such leaves and stems decay in a few years on modern marshes. Their preservation in tidal-flat mud above buried marsh soils implies that the change from marsh to tidal flat took a few years at most.

THOUGH HUNDREDS OF TREES succumbed to postearthquake submergence at Portage in 1964 (p. 14-15), some of those immersed in fresh water managed to live a few months beyond the March 27 earthquake. Their bark adjoins light-colored early wood from the 1964 growing season. Not imagining such survival, Atwater and Yamaguchi (1991) misinterpreted incomplete outer rings at Cascadia as evidence for sudden submergence during a growing season, between May and September. The trees in question are spruce that died from the Cascadia earthquake now dated to January 1700. As at Portage, some of the submerged spruce survived into the next growing season or later (Jacoby and others, 1995).

WIDE OUTER RINGS OF SPRUCE STUMP

One annual ring, composed of cells added early (light) and late (dark) in growing season, which runs May to September

1 cm

Final ring

Bay Center, Willapa Bay

LEAVES AND STEMS OF MARSH GRASS

Living tuft of the grass *Deschampsia caespitosa* on the high part of a tidal marsh.

Mud

Top of buried soil

15 cm

Fossil tuft interred by tidal-flat mud after subsiding circa A.D. 400 (p. 100). Plan view, mostly 1-2 cm above buried soil.

Niawiakum River, Willapa Bay

Sand sheets 地層中の砂層

Tsunamis overran newly dropped land along Cascadia's Pacific coast.

SIGNS OF CASCADIA TSUNAMIS

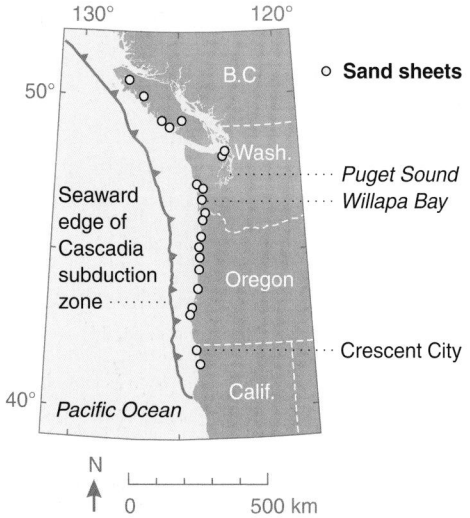

WHILE MAPPING Cascadia's signs of sudden subsidence, geologists in the 1980s and 1990s found associated evidence for tsunamis. That evidence consists of sand sheets beside bays and river mouths (dots on map, left). The sand came from the sea; it tapers inland and contains the microscopic siliceous shells of marine diatoms. Beside muddy bays the sand alternates with layers of mud (photos below) that probably settled out in lulls between individual waves in a tsunami wave train (modern example, opposite).

At most sites, the sand arrived just before tidal mud began covering a freshly subsided soil (cartoons below). Neither a storm nor a tsunami of remote origin explains this coincidence with subsidence. The simplest explanation is a tsunami from an earthquake in which a tectonic plate, in a seismic shift, abruptly displaces the sea while lowering the adjoining coast. The resulting tsunami then overruns the lowered land (cartoon, p. 10).

A TSUNAMI LAYS DOWN A SHEET OF SAND

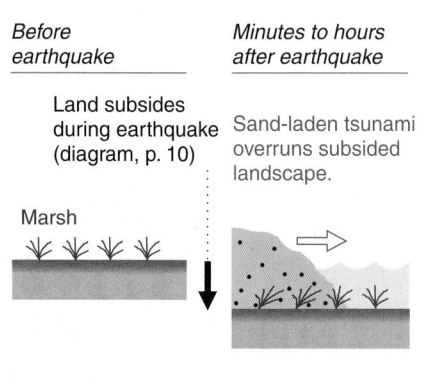

Before earthquake

Land subsides during earthquake (diagram, p. 10)

Marsh

Minutes to hours after earthquake

Sand-laden tsunami overruns subsided landscape.

Centuries after earthquake

Sand sheet

Detail below

Mud deposited by tides since 1700. Contains coarse silt but lacks sand except just above a subsided, buried soil. Mud thickness, 80 cm

Buried soil of marsh that subsided in 1700

SAND SHEETS from tsunamis of great Cascadia earthquakes have been identified along Cascadia's Pacific coast (compilation by Peters and others, 2003) and at northern Puget Sound (Williams and others, 2005). Some cover archaeological sites (p. 20-21) and the floors of coastal lakes (Hutchinson and others, 1997). Constituents include microscopic marine fossils (Hemphill-Haley, 1996). Sand sheets in British Columbia record Alaskan waves of 1964 in addition to the 1700 Cascadia event (Clague and others, 2000).

A SMALL TSUNAMI on April 25-26, 1992, in northern California, provides further evidence that the Cascadia subduction zone generates tsunamis of its own. The parent earthquake, of magnitude 7.1, probably broke the Cascadia plate boundary near its southern end. The tsunami crested 0.5 m above tides at Crescent City, where it lasted eight hours (Oppenheimer and others, 1993).

Inland ⟶

Grass bent inland by incoming tsunami

5 cm

TIDAL MUD

Grass tuft

BURIED MARSH SOIL

Sandy layer, each the likely record of an onrushing wave in the tsunami train that began the evening of January 26, 1700 (inferred timing, p. 42-43). The sandy layers alternate with mud that probably records the slack water of crested waves.

Niawiakum River, Willapa Bay (Oyster locality of Atwater and Hemphill-Haley, 1997)

Chilean counterparts

THE TSUNAMI associated with the giant 1960 Chile earthquake deposited sand in Chile. The deposit was noted soon afterward at several northern sites. Additional examples were documented decades later near Maullín, where tsunami sand had settled on subsided pastures.

Likewise in Japan, on the far side of the Pacific, the 1960 tsunami deposited sand onshore. For example, it coated plains beside Miyako Bay with alternating layers of sand and silt. The layers probably represent several of the dozens of 1960 tsunami waves recorded by the Miyako tide gauge. Those waves were numerous because, like the Cascadia tsunamis simulated on pages 37, 74-75, and 103, the 1960 Chile tsunami reflected off shorelines and resonated in bays.

In Chile

TSUNAMI DEPOSIT NEAR MAULLÍN

Tidal marsh

10 cm

Pasture soil　　**Sand sheet** deposited by 1960 tsunami; later etched by high tides

WAVE TRAIN AT TALCAHUANO TIDE GAUGE

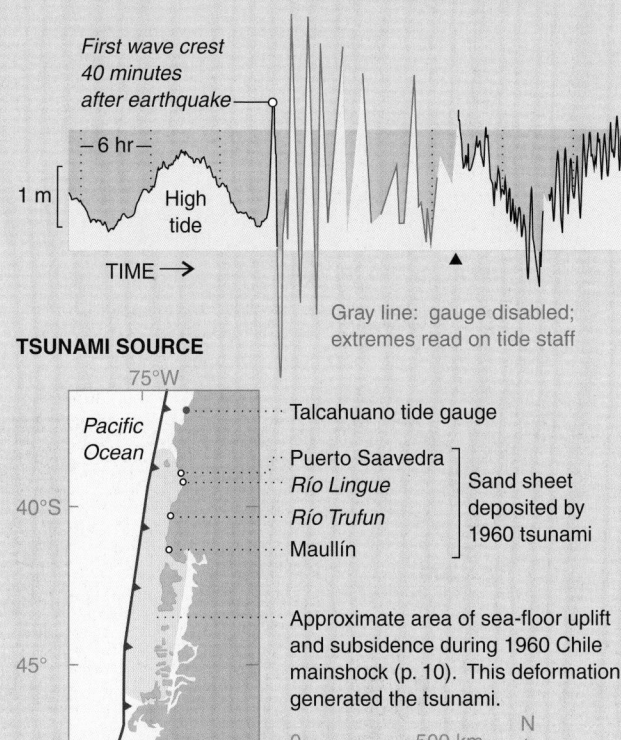

First wave crest 40 minutes after earthquake

— 6 hr —

1 m

High tide

TIME →

Gray line: gauge disabled; extremes read on tide staff

TSUNAMI SOURCE

75°W

Pacific Ocean

40°S

Talcahuano tide gauge

Puerto Saavedra
Río Lingue
Río Trufun
Maullín
] Sand sheet deposited by 1960 tsunami

45°

Approximate area of sea-floor uplift and subsidence during 1960 Chile mainshock (p. 10). This deformation generated the tsunami.

0　　500 km　　N

The parent earthquake took place 1710-1715 G.M.T., 22 May 1960

SAND SHEETS were noted by Wright and Mella (1963, p. 1371, 1372, 1389) and, near Maullín, by Cisternas and others (2005). Tide-gauge data redrawn from Sievers and others (1963, sheet 3). Tsunami source inferred from land-level changes mapped by Plafker and Savage (1970).

In Japan

TSUNAMI DEPOSIT BESIDE MIYAKO BAY

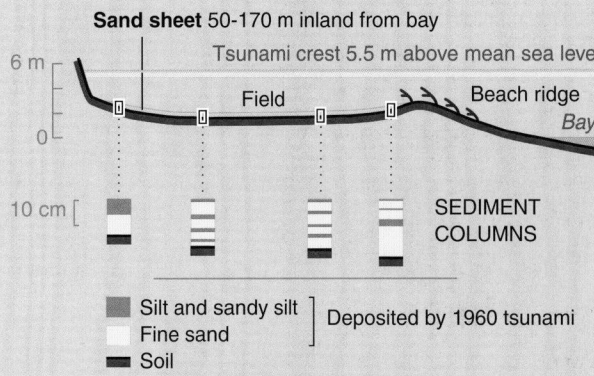

Sand sheet 50-170 m inland from bay

Tsunami crest 5.5 m above mean sea level

6 m

Field

Beach ridge

Bay

0

10 cm

SEDIMENT COLUMNS

■ Silt and sandy silt
□ Fine sand　　] Deposited by 1960 tsunami
■ Soil

WAVE TRAIN AT MIYAKO TIDE GAUGE

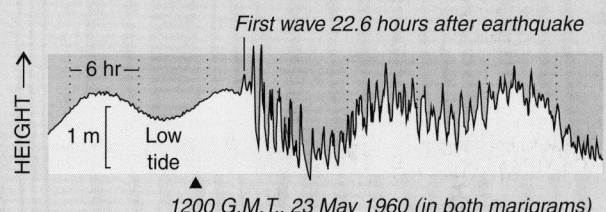

First wave 22.6 hours after earthquake

— 6 hr —

HEIGHT →

1 m

Low tide

1200 G.M.T., 23 May 1960 (in both marigrams)

TSUNAMI HEIGHTS BESIDE MIYAKO BAY (location, p. 51)

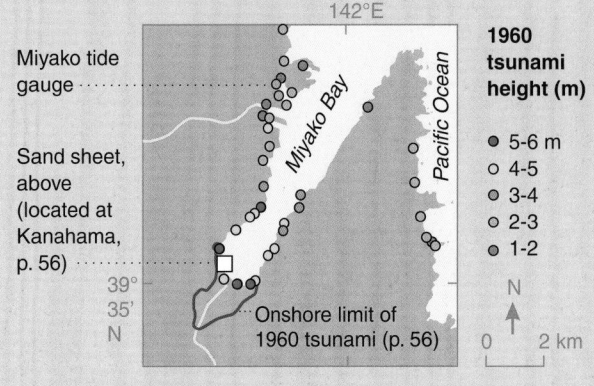

142°E

Miyako tide gauge

Sand sheet, above (located at Kanahama, p. 56)

39° 35' N

Miyako Bay

Pacific Ocean

Onshore limit of 1960 tsunami (p. 56)

1960 tsunami height (m)
● 5-6 m
○ 4-5
◍ 3-4
◐ 2-3
◕ 1-2

N

0　　2 km

G.M.T., Greenwich Mean Time

AT MIYAKO BAY the 1960 tsunami deposits contain microscopic marine fossils (Onuki and others, 1961) in addition to the multiple layers illustrated above (redrawn from Kitamura and others, 1961b). Details on the marigram, p. 46; sources for the mapped tsunami heights, p. 55.

In harm's way 危険地域

Earthquake-induced submergence ruined Cascadia campsites.

IN A YUROK MYTH recorded a century ago, Thunder wants people to have enough to eat. He thinks they will if prairies can be made into ocean. He asks Earthquake for help. Earthquake runs about, land sinks, and prairies become ocean teeming with salmon, seals, and whales.

In cruel reality, native people paid a price for whatever they gained when Cascadia's great earthquakes changed tidal prairies into shallow arms of the sea. First they faced horrific tsunamis, like the one implied by the story of a sea flood that swept canoes into trees (p. 12). Survivors then watched tides relentlessly cover their subsided, bayside fishing camps.

Several archaeological sites tell wordlessly of the waves and tides that overran them. Each lies buried beneath tidal mud. Some are also coated with tsunami sand. In the 1980s and 1990s, geologists noticed them while studying buried soils in the banks of tidal streams (examples below and opposite).

Most of the archaeological sites stand out for their broken stones. The estuaries' muddy banks rarely contain sand, much less stones. But native peoples brought in pebbles and cobbles. They baked them in hearths, then used them to heat water in woven baskets and wooden boxes. Thermally shocked, the stones shattered.

Did a tsunami put out the campfires? None of the identified sites tells a story so dramatic. And something else must have driven people from the fishing camp on the facing page. Probably it was abandoned a century or two before the earthquake that sank it.

FORMER FIRE PITS, OREGON

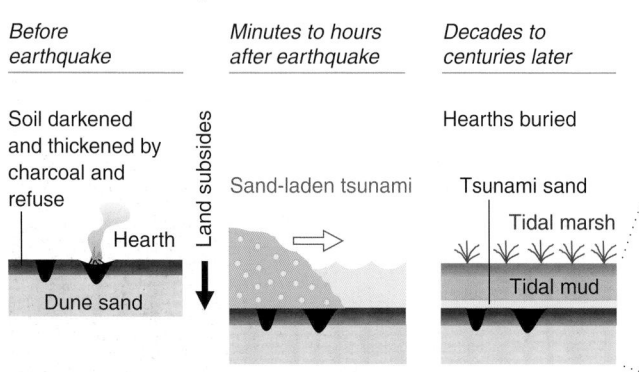

Before earthquake

Soil darkened and thickened by charcoal and refuse

Hearth

Dune sand

Land subsides

Minutes to hours after earthquake

Sand-laden tsunami

Decades to centuries later

Hearths buried

Tsunami sand

Tidal marsh

Tidal mud

Tsunami deposit

Tidal mud

10 cm

Hearths with charcoal and fire-modified rock

Soil amended by refuse

Dune sand

Salmon River, site 35LNC64

FIRE-MODIFIED ROCK, WASHINGTON

Entirely angular pieces

Unmodified pebbles also found at site

Broken pebbles and cobbles

1 cm

Niawiakum River, Willapa Bay, site 45PC102

Campsites ruined by a great Cascadia earthquake
○ Covered by tsunami sand and tidal mud
◎ Covered by tidal mud only
□ **Residence of Yukok teller** of "How the prairie became ocean"

B.C.

WASH.

···· *Willapa Bay*

···· *Nehalem River*

···· *Salmon River*

OREGON

— Espeu (at Big Lagoon)

Pacific Ocean

CALIF.

0   500 km

ANN OF ESPEU, of the Yurok tribe, recounted "How the prairie became ocean" to the ethnographer Alfred L. Kroeber (1876-1960) between 1900 and 1908 (Kroeber, 1976, p. 460).

EARTHQUAKE-INDUCED SUBSIDENCE has been inferred from burial of archaeological sites in Oregon (Minor and Grant, 1996; Hall and Radosevich, 1998) and Washington (Cole and others, 1996; source of rock photo).

A weaver's fate

WHAT BECAME OF THE MAKER of the woven object at right?

In 1991, before salvage by an archaeologist, the weaving protruded from an eroding tidal bank of Oregon's Nehalem River. It rested on the lowest centimeter of tidal mud that covers a buried marsh soil. Its radiocarbon age matches the time when the marsh changed into a tidal flat. Did the tsunami from a great Cascadia earthquake snatch the weaving from a coastal village? Did the weaver survive?

A PIECE OF THE WEAVING gave a radiocarbon age (173 ± 44 ^{14}C yr B.P.; GX-17835) statistically indistinguishable from the mean of 16 ages on stems and leaf bases found rooted in the soil and entombed in the overlying mud (179 ± 15 ^{14}C yr B.P.; Nelson and others, 1995; graphed on our p. 25).

] 1 cm

Tidal mud 1 m thick

View, above

Buried soil, exhumed

1 km north of Wheeler; "Downstream" site of Grant (1992).

ABANDONED FISHING CAMP, WASHINGTON

Surveyors on tidal marsh

View, below right

100 m to bank, below left

Weir

Weir

Tidal flat

Willapa River

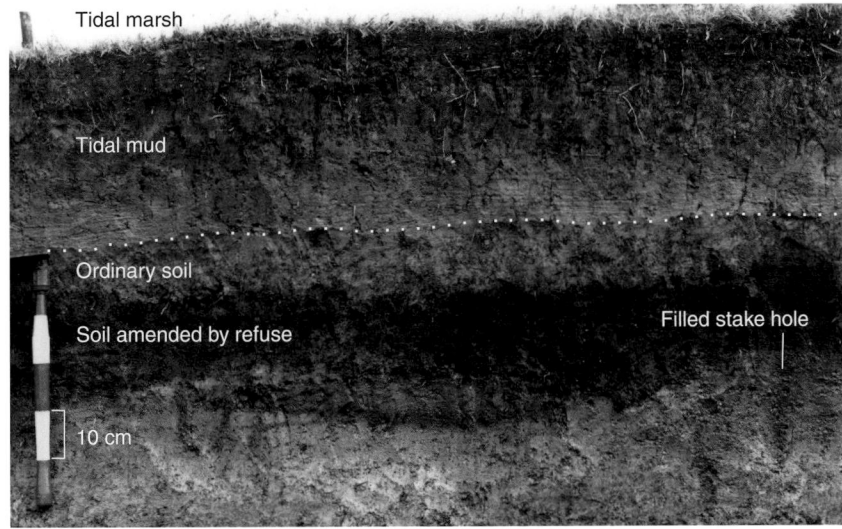

Tidal marsh

Tidal mud

Ordinary soil

Soil amended by refuse

Filled stake hole

10 cm

A FENCE IN WATER, a fishing weir blocks fish or directs them into traps. Dozens of prehistoric examples have been reported from the coast between southeast Alaska and northern California (Moss and Erlandson, 1998). The weirs above jut into the tidal Willapa River on the bank opposite downtown South Bend (site 45PC103, Atwater and Hemphill-Haley, 1997, p. 69-71 and figs. 29, 30).

Sea →

Two fishing weirs, exposed at very low tides at Willapa Bay, Washington, probably predate the 1700 Cascadia earthquake. A bark-bearing stave dates to 1400-1650. In the adjoining bank, archaeologically sterile soil records a time when the site lay abandoned. This soil separates a culturally darkened soil from the earthquake's signature—the abrupt upward change to distinctly laminated mud that postearthquake tides laid on subsided land.

Currents and cracks 水中土石流と液状化

Cascadia earthquakes avalanched sea-floor mud and quickened coastal sand.

DID THE EARTH QUAKE while land subsided and tsunamis began along Cascadia's Pacific coast? This key question went unanswered until the early 1990s, when two lines of evidence pointed to seismic shaking.

First, shaking offshore was shown to explain bottom-hugging muddy flows (turbidity currents) that repeatedly descended submarine channels (cartoon below).

Second and crucially, shaking onshore was found to have accompanied the coastal subsidence. The shaking liquefied loose, wet sand, turning it to quicksand. Water expelled from the liquefied sand erupted through cracks onto freshly subsided land (right). Today, these conduits are easy to spot because water plucks sand grains from the cracks more easily than it scours the sticky mud beside them.

EVIDENCE FOR SHAKING

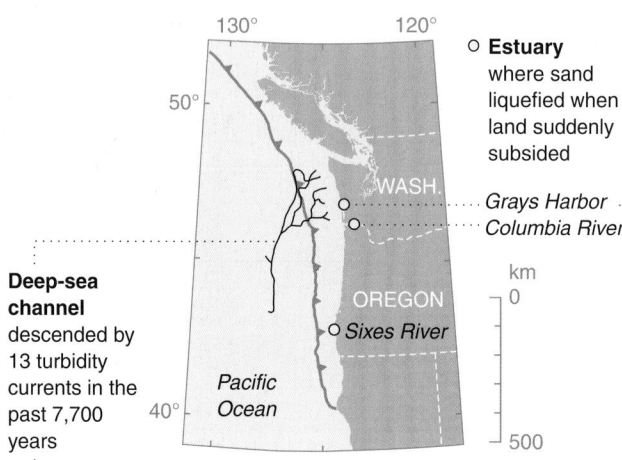

○ **Estuary** where sand liquefied when land suddenly subsided

Grays Harbor
Columbia River

WASH.

OREGON
○ *Sixes River*

Pacific Ocean

km
0

500

Deep-sea channel descended by 13 turbidity currents in the past 7,700 years

SHAKING LEAVES A DEEP-SEA DEPOSIT

1 **River** delivers sediment to the sea.
2 **Sediment** settles on the continental shelf.
3 **An earthquake** shakes the continental shelf and slope.
4 **Shaken sediment** descends submarine canyons as turbidity currents.
5 **Turbidity currents** merge where tributaries meet. Resulting deposits are visible in sediment cores.

Fault at plate boundary

ON TURBIDITE EVIDENCE for great Cascadia earthquakes, see Adams (1990; source of the above cartoon) and Goldfinger and others (2003).

LIQUEFACTION during the 1700 Cascadia earthquake produced sand dikes along the Columbia River. These were discovered in the early 1990s by Stephen Obermeier (Peterson, 1997; Obermeier and Dickenson, 2000). Probably correlative intrusions were later found at Grays Harbor (right) and at Sixes River, Oregon (Kelsey and others, 2002, p. 310-312). The 1964 Alaska earthquake generated dikes near Portage (cracks in photo, p. 14; Walsh and others, 1995).

SHAKING YIELDS A SAND-FILLED CRACK

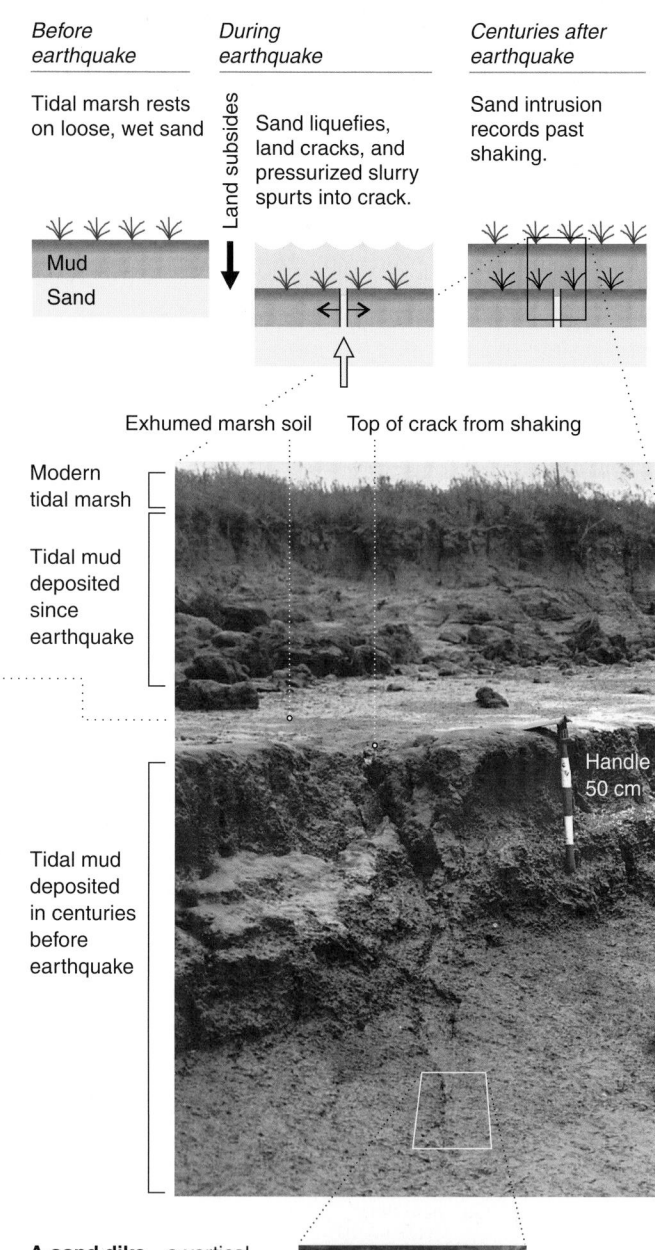

Before earthquake

Tidal marsh rests on loose, wet sand

Mud
Sand

During earthquake

Land subsides

Sand liquefies, land cracks, and pressurized slurry spurts into crack.

Centuries after earthquake

Sand intrusion records past shaking.

Exhumed marsh soil   Top of crack from shaking

Modern tidal marsh

Tidal mud deposited since earthquake

Tidal mud deposited in centuries before earthquake

Handle 50 cm

A sand dike—a vertical sand-filled crack—rises to the top of a marsh soil, seen above in a natural low-tide outcrop. In the plan view at right, the dike cuts sharply across the mud.

Johns River, Grays Harbor (p. 103). In lower photo, the grooves in the mud are from scraping tool.

5 cm

Mud   Sand   Mud

Strength of shaking

TO DESIGN A SCHOOL to withstand a great Cascadia earthquake, an engineer needs to know what shaking to expect. Researchers have sought guidance from records of past shaking at Cascadia, thus far with little success.

To estimate ancient ground motions, a logical first step is to identify sand that an earthquake liquefied. The sand's resistance to liquefaction can then be measured, and the results compared with those from sand that did or did not liquefy at known levels of shaking.

However, sand that liquefies can look just the same as it did before. It can retain its original sedimentary layers after expelling the water that drives intrusions. In the photo below, a sill and offshooting dikes show that sand liquefied somewhere below them. How can that source sand be identified, so that its resistance to liquefaction can guide the design of schools that resist earthquakes?

SEDIMENT SLICER BESIDE COLUMBIA RIVER

Photo by Bill Wagner, Longview Daily News, September 2000

INTRUSIONS IN VERTICAL SLICE

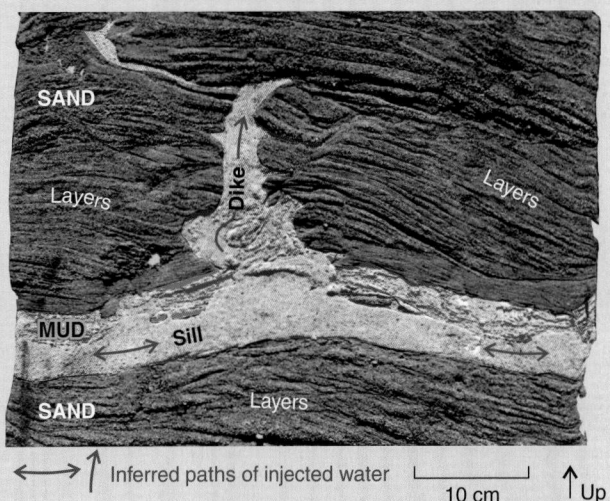

Inferred paths of injected water   10 cm   ↑ Up

SCHEMATIC DISTRIBUTION OF INTRUSIONS

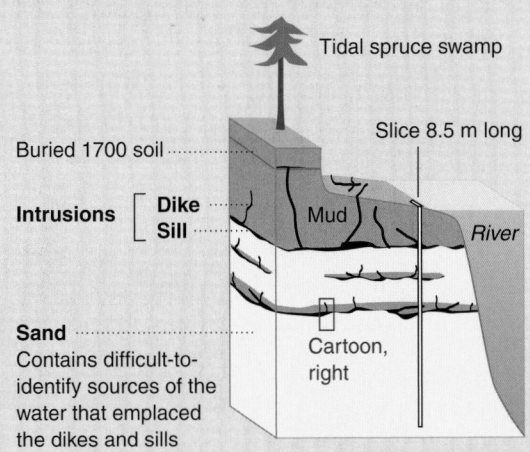

PEAK SURFACE ACCELERATIONS of 0.15-0.35 g were inferred from the localized absence of near-surface sand dikes along the Columbia River 35-60 km inland from the Pacific coast (Obermeier and Dickenson, 2000). Deeper liquefaction features, like those at upper right, cast doubt on these proposed upper bounds (Atwater and others, 2001a; Takada and Atwater, 2004).

DURING LIQUEFACTION, sand sheared by shaking loses strength through an increase in pore-water pressure that decreases grain-to-grain contact. Partial collapse of the grain structure then drives much of the water out. However, where the expelled water escapes diffusely, the primary sedimentary layers in liquefied sand can remain nearly intact (Lowe, 1975; Owen, 1987; Liu and Qiao, 1984).

ADDED TINTS highlight the intrusions and mud bed in the slice photo.

INFERRED EMPLACEMENT OF INTRUSIONS

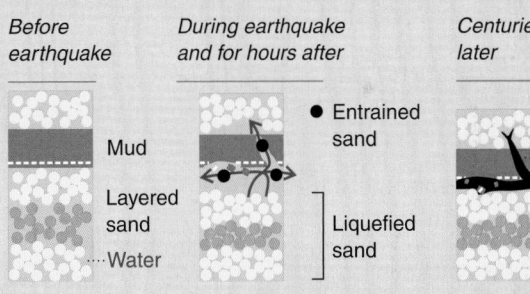

Before earthquake

During earthquake and for hours after

Centuries later

Loose sand, thinly layered, is interbedded with mud. Water occupies the space between sand grains.

Sand liquefies; grains lose contact with one another. They then settle into a more compact arrangement. The compaction drives out water, which initially percolates too slowly to erase sedimentary layers. However, where dammed by a mud bed, the water ponds and begins streaming sideways, moving grains against gravity. Locally it erodes the mud from beneath and breaks through the mud into sand above.

Intrusions provide the sole conspicuous sign that liquefaction occurred meters beneath them. Though more compact than it had been before the earthquake, the layered sand that liquefied retains most of its original sedimentary layering.

Magnitude 9? マグニチュード９？

Geologists reach an impasse on Cascadia's potential for a giant earthquake.

MAXIMUM EARTHQUAKE SIZE remained a big unknown for Cascadia through the early 1990s.

By then, geologists had identified signs of earthquake-induced subsidence, and attending tsunamis, at estuaries from southern British Columbia to northern California. They knew that nearly all sites had dropped most recently within the past 400 or 500 years, and that the southern Washington coast subsided in the decades after A.D. 1680 (box, below).

These findings spurred a radiocarbon experiment designed to detect coastwise differences—if any—in the time of earthquake-induced subsidence. Any such differences would limit earthquake size by limiting fault-rupture length.

The experiment ended up denying neither the giant-earthquake hypothesis nor its serial alternative (opposite). The most exact of the ages show that trees nearly 700 km apart, in southern Washington and northern California, died from earthquake-induced subsidence during the same few decades. Either a single giant earthquake or a swift series of merely great earthquakes could have done the job.

But the experiment succeeded in narrowing the time window for Cascadia's most recent giant earthquake, or great-earthquake series, to the period 1695-1720. And unbeknown to the experimenters, an orphan tsunami in 1700 had long been puzzling earthquake historians in Japan.

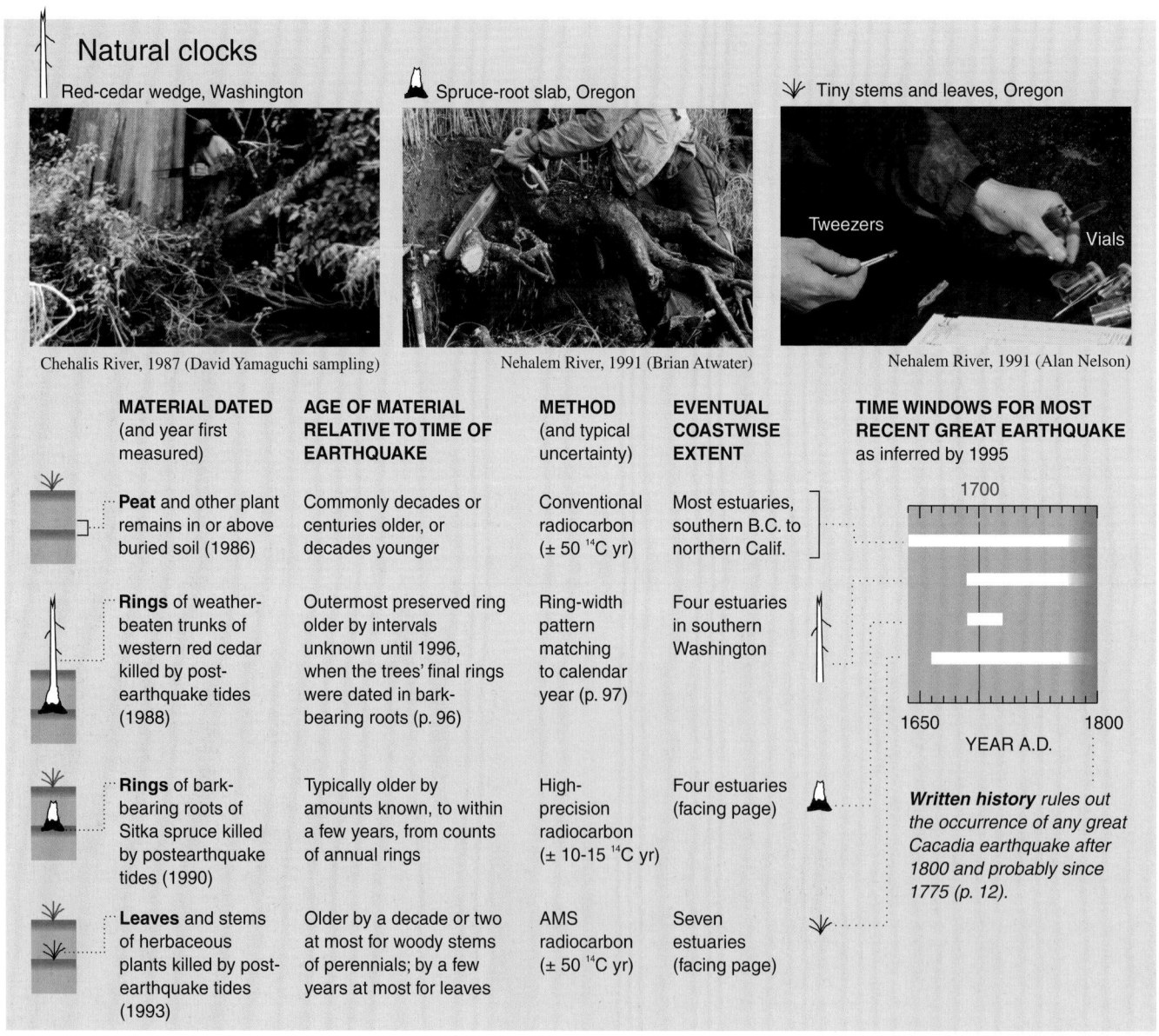

Natural clocks

Red-cedar wedge, Washington

Spruce-root slab, Oregon

Tiny stems and leaves, Oregon

Tweezers   Vials

Chehalis River, 1987 (David Yamaguchi sampling)

Nehalem River, 1991 (Brian Atwater)

Nehalem River, 1991 (Alan Nelson)

MATERIAL DATED (and year first measured)	AGE OF MATERIAL RELATIVE TO TIME OF EARTHQUAKE	METHOD (and typical uncertainty)	EVENTUAL COASTWISE EXTENT	TIME WINDOWS FOR MOST RECENT GREAT EARTHQUAKE as inferred by 1995
Peat and other plant remains in or above buried soil (1986)	Commonly decades or centuries older, or decades younger	Conventional radiocarbon (± 50 ^{14}C yr)	Most estuaries, southern B.C. to northern Calif.	
Rings of weather-beaten trunks of western red cedar killed by post-earthquake tides (1988)	Outermost preserved ring older by intervals unknown until 1996, when the trees' final rings were dated in bark-bearing roots (p. 96)	Ring-width pattern matching to calendar year (p. 97)	Four estuaries in southern Washington	
Rings of bark-bearing roots of Sitka spruce killed by postearthquake tides (1990)	Typically older by amounts known, to within a few years, from counts of annual rings	High-precision radiocarbon (± 10-15 ^{14}C yr)	Four estuaries (facing page)	
Leaves and stems of herbaceous plants killed by post-earthquake tides (1993)	Older by a decade or two at most for woody stems of perennials; by a few years at most for leaves	AMS radiocarbon (± 50 ^{14}C yr)	Seven estuaries (facing page)	

1700

1650   1800
YEAR A.D.

Written history rules out the occurrence of any great Cacadia earthquake after 1800 and probably since 1775 (p. 12).

For examples of differences in geological and analytical precision, see Atwater and Hemphill-Haley (1997, p. 84 [soil Y] versus p. 89).

±, standard deviation reported by lab. ^{14}C yr, radiocarbon years (graphed, opposite). AMS, accelerator mass-spectrometry, used to date small samples.

HYPOTHESES: ALL AT ONCE OR A PIECE AT A TIME?

Giant earthquake, M 9 | **Series of lesser events, M 8+**

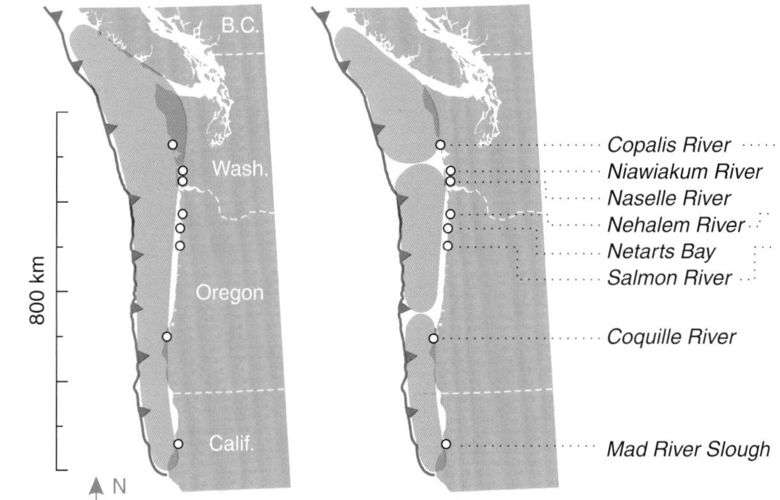

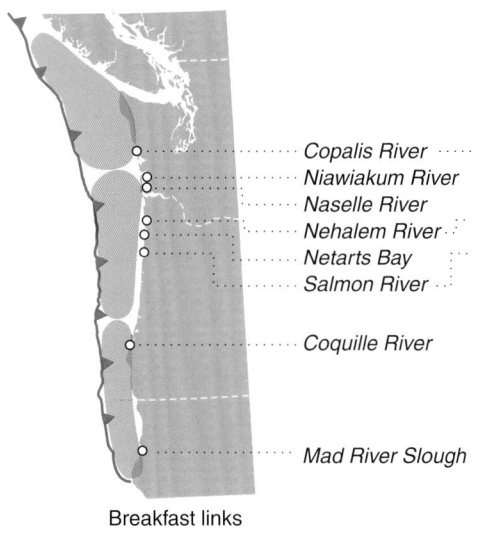

B.C.
Wash.
Oregon
Calif.
800 km
N
Copalis River
Niawiakum River
Naselle River
Nehalem River
Netarts Bay
Salmon River
Coquille River
Mad River Slough

Dinner sausage | Breakfast links

SKEPTICS of the giant-earthquake hypothesis felt that a series of lesser earthquakes is "more consistent with observations at other subduction zones" (McCaffrey and Goldfinger, 1995).

THE RADIOCARBON AGES at right were reported by Atwater and others (1991) and Nelson and others (1995). Each black error bar spans the ranges of individual ages. Each green-coded plant yielded one or two individual ages.

INCONCLUSIVE TEST: DEATH AT THE SAME TIME?

The remains of earthquake-killed plants yielded ages that are neither statistically different nor necessarily the same within each of the three groups color-coded below.

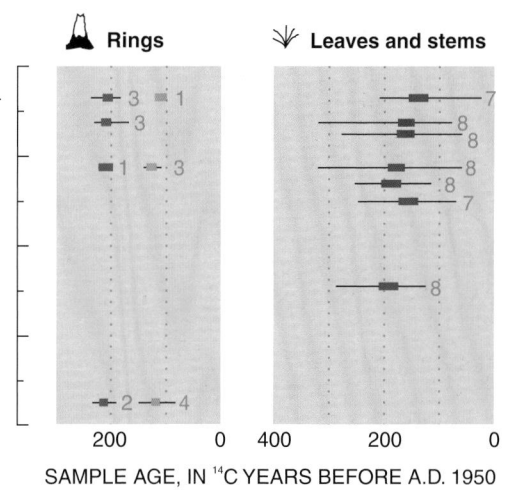

Rings
Leaves and stems
3
1
3
1
3
2
4
7
8
8
8
8
7
8
200
0
400
200
0
SAMPLE AGE, IN ¹⁴C YEARS BEFORE A.D. 1950

Mean ages, ± 1 standard error. Wood coded by mean number of ring years before tree death: 40 ▬ 5 ▬

Range of sample ages

Number of plants dated

Uncommonly exact

RADIOCARBON AGES rarely pin down the time of an event. To narrow the time of Cascadia's most recent giant earthquake (or serial great earthquakes) to 1695-1720, isotopists and geologists pushed radiocarbon precision to its limits. They took advantage of quirks in the radiocarbon timescale, and they maximized geological and analytical precision in sampling and measurement.

Radiocarbon time has been called rubberband time. It stretches and shrinks because radioactive carbon is produced in Earth's atmosphere by cosmic rays whose flux waxes and wanes. Trees use radiocarbon as part of the atmospheric carbon dioxide from which they make their annual rings. Tree rings thus yield radiocarbon ages that wiggle away from straight-line equivalence of radiocarbon and calendar time. One of the tallest jags spans most of the century before A.D. 1700 (graph at right).

The dating to 1695-1720 relied on finding this tall jag in the annual rings of earthquake-killed spruce. Ring counts adjust for the time lag between the dated rings and the tree-killing earthquake. The radiocarbon ages themselves, like those that define the calibration curve, were measured at uncommon precision on cellulose whose carbon the trees took from the atmosphere by photosynthesis shortly before the dated rings formed.

Sample selection was guided by red-cedar evidence that the earthquake followed the 1680s (graph, left). This evidence, honed in the late 1990s (p. 96-97), would strengthen Cascadia's link to a tsunami in Japan.

CALIBRATION CURVE
Relates radiocarbon ages (above) to calendar dates. Derived from tree rings of known date. Wood of unknown date can be matched to a wiggle by the radiocarbon dating of rings that are known numbers of ring years apart.

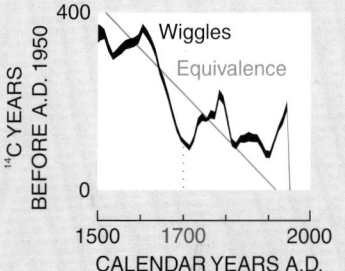

400
0
¹⁴C YEARS BEFORE A.D. 1950
Wiggles
Equivalence
1500
1700
2000
CALENDAR YEARS A.D.

GEOLOGICAL PRECISION
Tree rings give exact difference between age of dated material and time of earthquake.

Dated carbon incorporated by photosynthesis.
Rings mark intervening years.
Earth-quake kills tree.
TIME

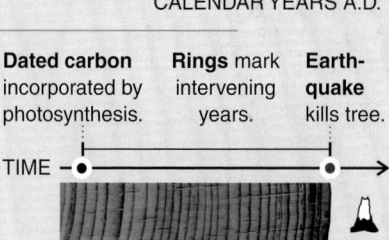

ANALYTICAL PRECISION
Analyzed for weeks in shielded counters, tree rings can be dated with uncertainties of 10-20 ^{14}C years. Cellulose, the skeleton of wood, is first extracted to limit the dated carbon to the years the rings formed.

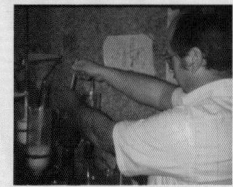

Seattle, 1996 (Phil Wilkinson)

THE RANGE 1695-1720 contains the 95-percent confidence interval; at most there is a 1-in-20 chance that the dated event occurred outside the range. Stuiver and others (1991) reviewed controls on the radiocarbon timescale. The calibration curve is from Stuiver and others (1998).

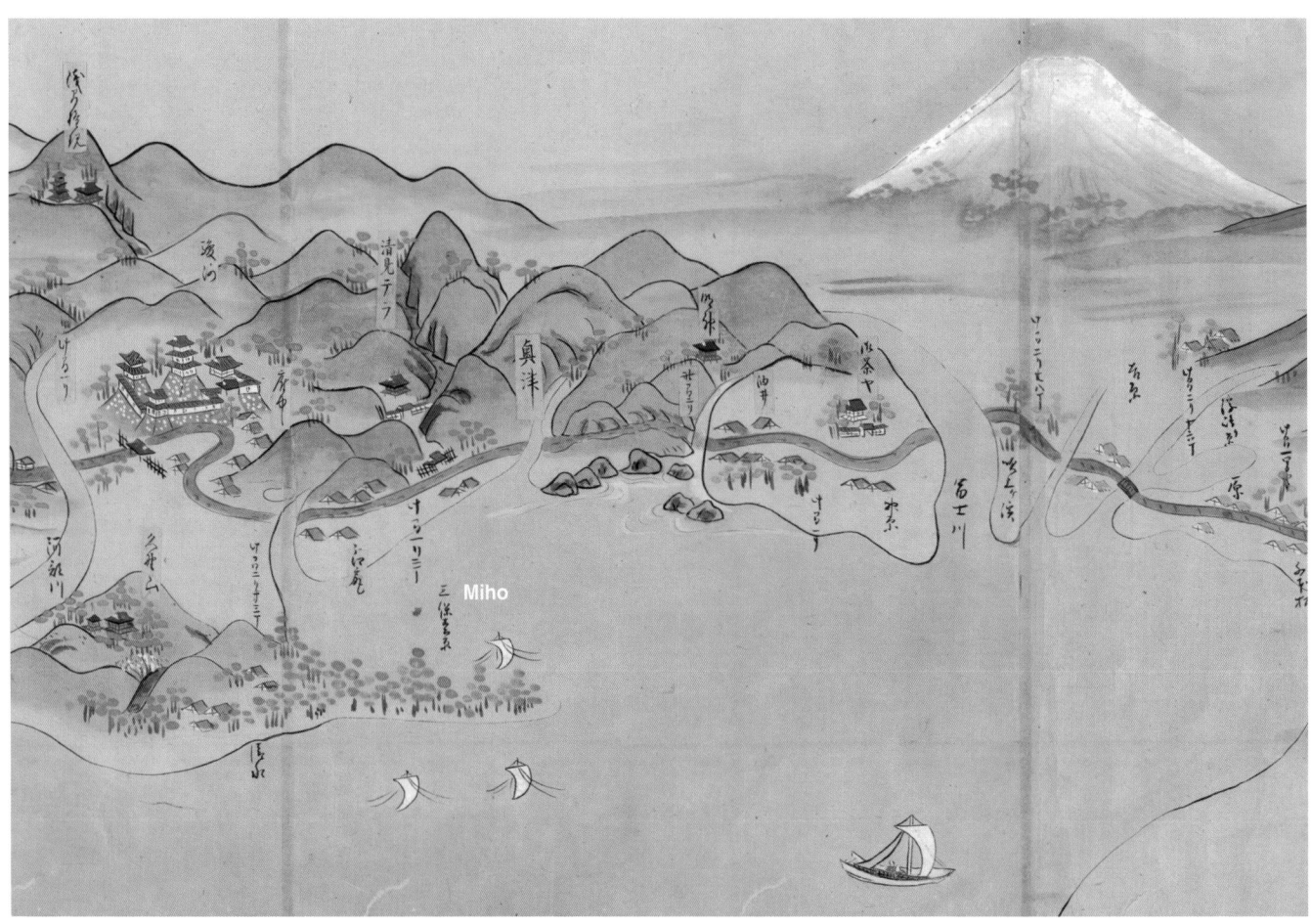

On the scenic spit at Miho, a village leader puzzled over a train of waves in January 1700 (p. 40, 78-79).

VIEW OF MIHO, and of the snowy cone of Mount Fuji, is from "Tōkaidō narabini saigoku dōchū ezu," 1687 (details, p. 76). Courtesy of East Asian Library, University of California, Berkeley.

CASTLE is the Sumpu retirement home of Tokugawa Ieyasu (1542-1616), first in the line of shoguns who ruled Japan from Edo (now Tokyo) in 1603-1867. Enlarged view, p. 41.

DIARY EXCERPT from "Miho-mura yōji oboe," p. 78, columns 10-11.

tsunami
Tsunami

nado to
and such

mōsu koto
what is called,

kayō no gi ni
such a thing

sōrō ya
could it
be?

Part 2
The orphan tsunami　みなしご津波

A PACIFIC TSUNAMI flooded Japanese shores in January 1700. The waters drove villagers to high ground, damaged salt kilns and fishing shacks, drowned paddies and crops, ascended a castle moat, entered a government storehouse, washed away more than dozen buildings, and spread flames that consumed twenty more. Return flows contributed to a nautical accident that sank tons of rice and killed two sailors. Samurai magistrates issued rice to afflicted villagers and requested lumber for those left homeless. A village headman received no advance warning from an earthquake; he wondered what to call the waves (quote, opposite).

These glimpses of the 1700 tsunami in Japan survive in old documents written by samurai, merchants, and peasants. Several generations of Japanese researchers have combed such documents to learn about historical earthquakes and tsunamis. In 1943 an earthquake historian included two accounts of the flooding of 1700 in an anthology of old Japanese accounts of earthquakes and related phenomena. By the early 1990s the event had become Japan's best-documented tsunami of unknown origin.

Part 2 of this book contains a chapter for each of six main Japanese villages or towns from which the 1700 tsunami is known. Each chapter begins with a summary of main points, a geographical and historical introduction, and the content of the tsunami account itself. Other parts of the chapters explore related human and natural history. Concluding estimates of tsunami height reappear in Part 3 as clues for defining hazards in North America.

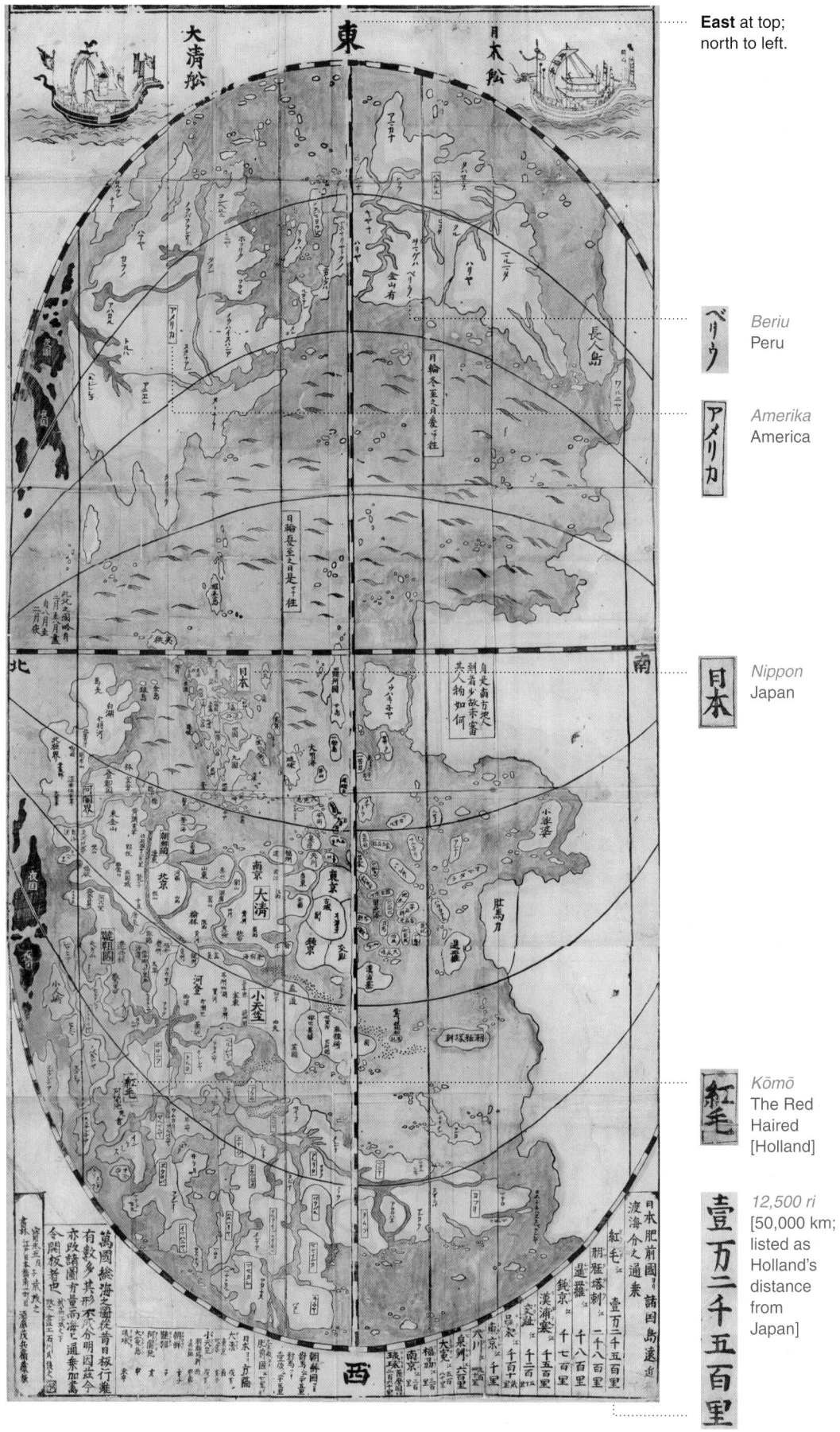

East at top;
north to left.

Beriu
Peru

Amerika
America

Nippon
Japan

Kōmō
The Red
Haired
[Holland]

12,500 ri
[50,000 km;
listed as
Holland's
distance
from
Japan]

Literate hosts 文字を使える人たち

In Japan, the 1700 tsunami reached a society ready to write about it.

THE YEAR 1700, though almost a century earlier than the first written records from northwestern North America, comes late in the written history of Japan. The year belongs, moreover, to an era of Japanese stability, bureaucracy, and literacy that promoted record-keeping.

That era began with national pacification early in the 17th century. By 1700, the country had known almost a century of peace for the first time in 500 years. Many in its military class were making their livings as bureaucrats. Samurai did paperwork for the Tokugawa shogun, the national leader in Edo (now Tokyo). They also administered the hinterlands as vassals of regional land barons, the daimyo.

Reading and writing extended beyond this ruling elite to commoners urban and rural. Booksellers offered poetry, short stories, cookbooks, farm manuals, and children's textbooks. Merchants tracked goods and services in an economy driven by bustling cities. Peasants prepared documents for villages they headed.

The accounts of the 1700 tsunami accordingly come from representatives of three social classes. The writers were military men employed by daimyo domains (p. 44, 70), merchants in business and local goverment (p. 53, 85), and peasants serving as village officials (p. 70, 77).

PERIOD MAPS open windows into the society in which those samurai, merchants, and peasants wrote. Such maps help introduce each of the six chapters in this part of the book. As a further introduction to a bygone time and place, consider the career of a commercial mapmaker and two of the products he sold: a decorative map of the world (opposite) and a travel map of Japan (overleaf).

Ishikawa Tomonobu wrote and drew in the decades around 1700. In addition to making maps, he illustrated calendars and novellas, composed linked-verse poetry and humorous fiction, and published travel guides and courtesan evaluations. Like many of his contemporaries, including the short-story writer Ihara Saikaku and the playwright Chikamatsu Monzaemon (p. 63), Ishikawa worked in a tradition, *ukiyo*, or floating world, that focused on daily life and its fleeting pleasures.

Ishikawa's world map, descended from 16th-century European compilations, was modeled on 17th-century Japanese surveyors' certificates. The map served as an interior decoration hung lengthwise, east to the top. A companion sheet contained portraits of the world's peoples.

The map depicts an ocean between the Japanese islands from the Americas. Japanese phonetic symbols identify America and Peru. Chinese characters for "The Red Haired" denote Holland, Japan's sole European trading partner between 1639 and 1854.

The travel map, "Nihon kaisan chōriku zu," depicts Japan "sea to mountains." Ishikawa issued its first edition in 1691, woodblock printed and hand colored. The version on the overleaf dates from 1694.

Ishikawa makes "Nihon kaisan" useful to the traveler by fitting his subject into a rectangular format and by filling margins with tourist information. Marginal tables give travel distances, domestic and international. Half the domestic table gives distances by land; the other half, distances by sea. Additional tables name the most important shrine in each county of each province. The lower left corner of the map provides an almanac on solstices, equinoxes, phases of the moon, and tides. Above it, signs of the Chinese zodiac denote twelve compass directions (p. 43).

Frequent travelers in Ishikawa's Japan included daimyo and their entourages, who journeyed to Edo every year or two for required attendance upon the shogun. A square or circle on the travel map represents each daimyo domain. An adjoining label gives a measure of daimyo status—the domain's official valuation in terms of rice yield (p. 71)—and the name of the daimyo himself.

The ukiyo artist further depicts cities, castles, highways, fishermen, merchant marines, and urban samurai. Roofs represent the urban sprawl of the shogun's capital, Edo, its population soon to surpass one million. The Tōkaidō, or Eastern Sea Road, wends its way toward Kyoto, the imperial capital since A.D. 794. Fifty-three way stations await travelers seeking overnight accomodations.

Just off the Tōkaidō, the pines of Miho beckon from a floating-world island. On a peninsula rendered more accurately on page 26, in a village of 300 peasants, a farmer or fisherman will soon write the most vivid and inquisitive of Japan's accounts of the orphan tsunami of 1700 (p. 78-79).

ON JAPAN under the Tokugawa shoguns, see Totman (1993). Chibbett (1977, p. 123) reviews the origins of *ukiyo-e,* floating-world pictures and paintings.

ISHIKAWA TOMONOBU (or Ishikawa Ryūsen) is profiled in a Japanese literary encyclopedia, Nihon Koten Bungaku Daijiten Henshū I'inkai, (1983, p. 129).

ISHIKAWA'S WORLD MAP, "Bankoku sōkaizu" ("General world map"), is reproduced courtesy of the East Asian Library, University of California, Berkeley. Unno (1994, p. 404-409) traces its origins to 16th-century Chinese copies of European maps. Those copies, and Ishikawa's version as well, retain 16th-century European speculation on an enormous southern continent and on the shape of western North America. The Chinese copies were made under the direction of Matteo Ricci (1552-1610), a Jesuit missionary. Examples soon reached Japan; by 1605, Jesuits in Kyoto were using Ricci maps to teach geography. A Japanese adaptation of the Ricci model appeared by 1645. In his version, Ishikawa revised Ricci's Asian geography and stylized other parts of the map. His 12,500-ri distance to Holland, listed also on the tourist map overleaf (p. 30), exceeds Earth's circumference (40,074 km at the equator) if his *ri* equals 3.93 km (the conventional conversion; Nelson and Haig, 1997, p. 1268).

Tourist map, 1694

南
京
大坂御城
河内
和泉
大和
伊賀
伊勢
紀伊
尾張
美濃
参河

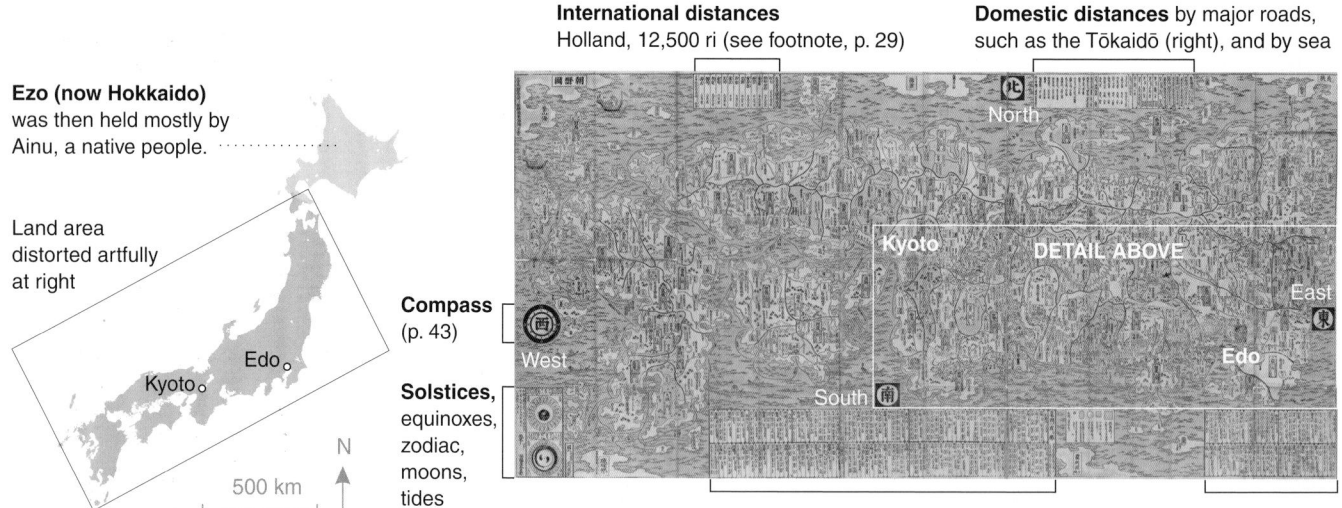

Ezo (now Hokkaido) was then held mostly by Ainu, a native people.
Land area distorted artfully at right
Edo
Kyoto
N
500 km

International distances
Holland, 12,500 ri (see footnote, p. 29)
Domestic distances by major roads, such as the Tōkaidō (right), and by sea
North
Kyoto
DETAIL ABOVE
East
Edo
West
South
Compass (p. 43)
Solstices, equinoxes, zodiac, moons, tides
Temples and shrines listed by kuni (ancient province) and gun (county)

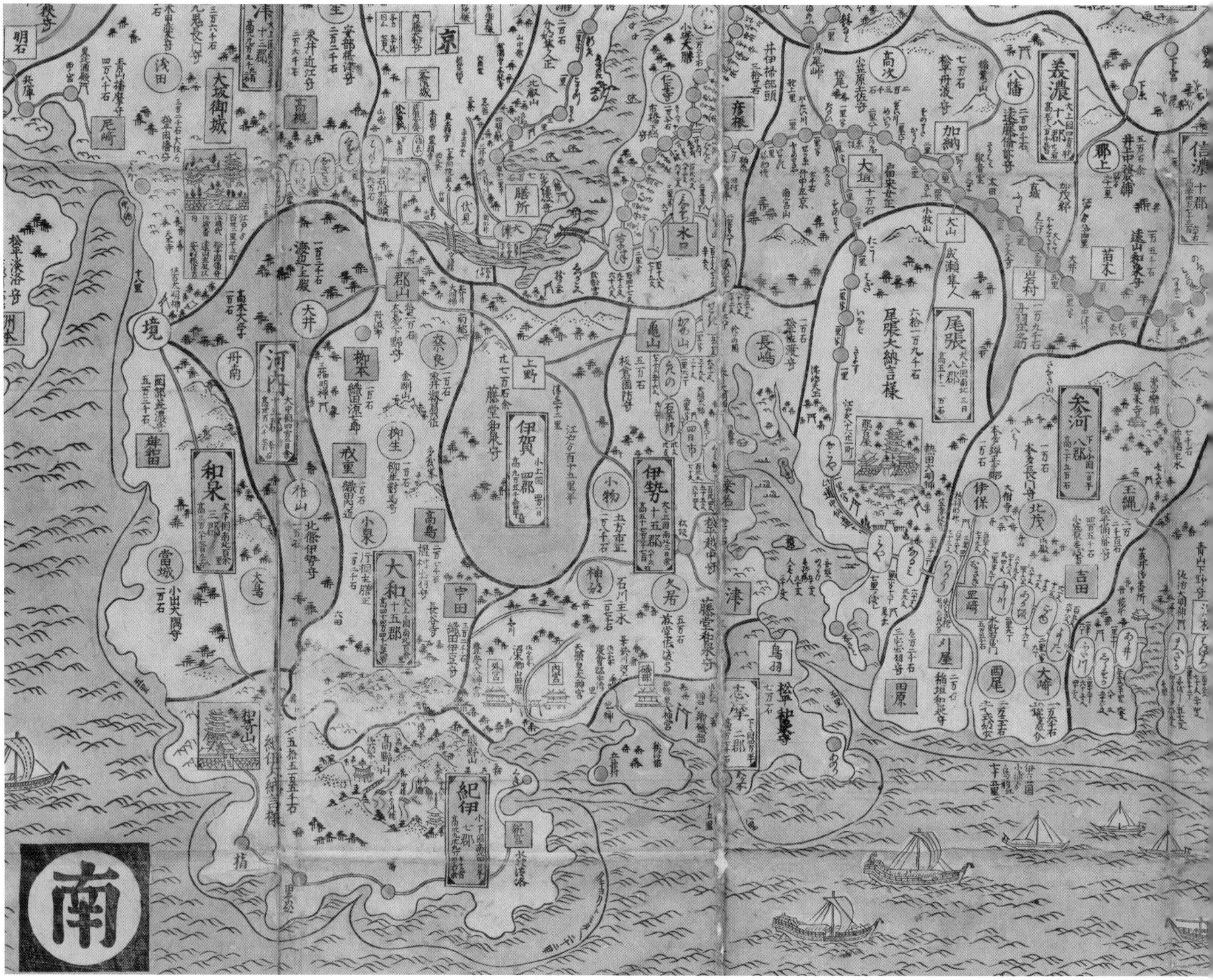

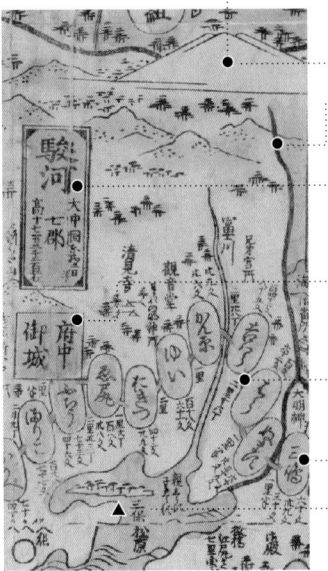

Mount Fuji

Province boundary
(shown more exactly on
official map, next page).
Suruga 駿河 province
contained seven *gun* 七郡,
or counties.

Sumpu castle, site of
earliest known writing of
"tsunami" (p. 41)

Tōkaidō, the "Eastern Sea
Road," connected imperial
Kyoto and shogunal Edo.
One of 53 way stations.

Miho 三保, source of an
account of the 1700
tsunami (p. 76-79)

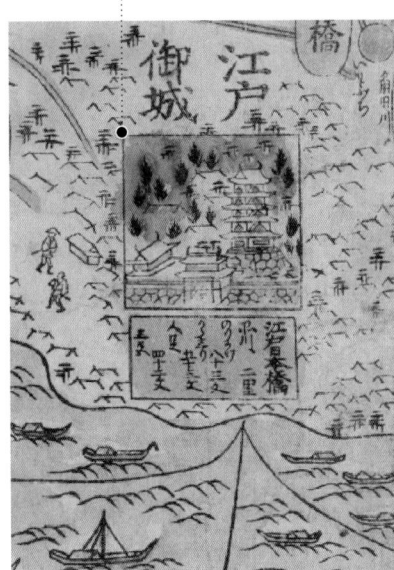

Edo 江戸,
the Tokugawa shoguns'
capital. Population
approaching one million
in 1700 (p. 61). Became
Tokyo in 1868.

"NIHON KAISAN
CHŌRIKU ZU," by Ishikawa
Tomonobu (p. 29), 1694
edition (Akioka, 1997, p.
214), fills a sheet nearly 1.7 m
by 1.2 m. Walter (1994, p.
194) likens the geographic
distortion to that in a subway
map. Courtesy of East Asian
Library, University of
California, Berkeley. More
excerpts, p. 43, 70-72.

Wetted places 浸水した地域

The orphan tsunami flooded sites along nearly 1000 kilometers of Japan's Pacific coast.

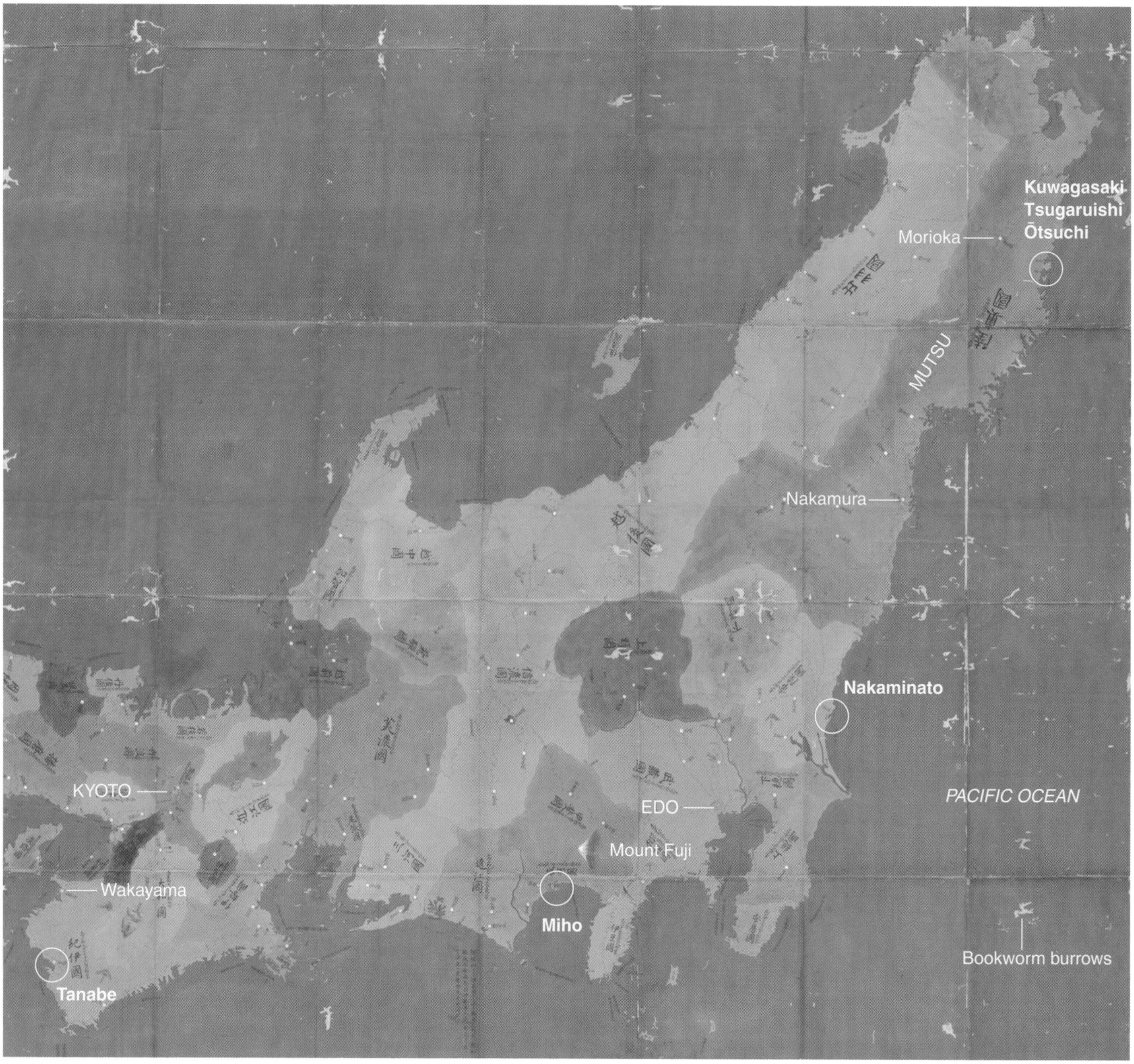

Kuwagasaki
Tsugaruishi
Ōtsuchi

Morioka

MUTSU

Nakamura

Nakaminato

KYOTO

PACIFIC OCEAN

EDO

Mount Fuji

Wakayama

Miho

Bookworm burrows

Tanabe

Part 2 of this book follows the January 1700 tsunami southwestward along Japan's Pacific coast from Kuwagasaki to Tanabe.

1702

THE TOKUGAWA SHOGUNATE, which ruled Japan from 1603 to 1867, ordered the entire country mapped five times. The fourth such mapping, in 1697-1702, produced a sheet for each of Japan's 83 ancient provinces at 1:21,600 scale (slightly larger than the standard scale of today's 7.5-minute topographic maps in the United States). From these detailed maps the shogunate compiled "Genroku Nihon sōzu," above, a map of all Japan in the Genroku era (Unno, 1994, p. 397, 472).

THE MAP DEPICTS the provinces in various earth tones. Bookworms (p. 87) made the burrows that unfold in symmetrical pairs.

FROM THE COLLECTION of Ashida Koreto (1877-1960); "Genroku Nihon sōzu" is map 09-110 of Ashida Bunko Hensan I'inkai (2004, p. 154). The entire map, now in two pieces, spans 3.1 by 4.4 meters. Courtesy of the library of Meiji University, Tokyo.

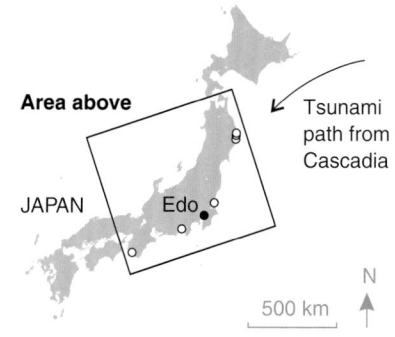

Area above

Tsunami path from Cascadia

JAPAN

Edo

N

500 km

PLACES FLOODED by the 1700 tsunami in Japan include Kuwagasaki, Tsugaruishi, Ōtsuchi, Miho, and Tanabe. Some of the accounts mention damage in additional villages. In one account, the tsunami takes the form of rough seas that initiate a nautical accident near Nakaminato. The writers represent three of their society's four main classes: the *bushi*, or samurai; farmers and other peasants; and merchants (p. 53).

The main accounts grace the next two pages. We parse them, from north to south, in the six chapters that follow.

PLACE	OTHER SITES where account was written or where tsunami entered	WRITERS	LOSSES
土 samurai — 農 peasant — 商 merchant			⌒ buildings — ‖‖ fields or crops — ⊿ salt kilns
Kuwagasaki *p. 36-49* ↳ Adjoined Miyako, where Morioka-han had a district office. Nearly 300 houses	The tsunami account originated in Miyako. It was delivered inland to Morioka, where it entered administrative records of Morioka-han.	土 Magistrates in Miyako and scribes in Morioka castle	⌒ 13 houses destroyed by flooding, 20 more by a concurrent fire
Tsugaruishi *p. 50-57* ● Along a farmed plain and a river known for crook-nose salmon. Nearly 200 houses	The writer describes losses along the nearby bayshore and mentions, as hearsay, the flooding and fire in Kuwagasaki.	商 Family that later purchased samurai status	⌒ Houses destroyed by flooding along bayshore near Tsugaruishi
Ōtsuchi *p. 58-65* ↳ Like Miyako, headquarters of an administrative district of Morioka-han	The tsunami account originated in Ōtsuchi. A summary survived there, as do details in administrative records in Morioka. Losses were said to have been reported to Edo.	土 Magistrates in Ōtsuchi and scribes in Morioka castle	*Damaged:* ⌒ 2 ⊿ 2 ‖‖ paddies and fields
Nakaminato *p. 66-75* ↳ Transferred cargo between seagoing ships and river boats that plied inland waterways to Edo	The account focuses on a shipwreck in rocks offshore of Isohama village, nearby. The cargo originated in Nakamura-han. Officials of Mito-han investigated.	農 The boat's crew and officials of Isohama village 土 Officials of Mito-han	Two sailors killed and nearly 30 tons of rice sunk in an accident caused mainly by a storm
Miho *p. 76-83* ● Picturesque place near the Tōkaidō. Population 300	The account remained in Miho, where it was later included in an anthology of headmen's writings.	農 Village headman	No damage reported
Tanabe *p. 84-92* ☐ Capitol of a sector of Wakayama-han. Population no less than 2600	Farming or fishing settlements near Tanabe: Atonoura, Mera, Mikonohama, and Shinjō	商 Mayor of Tanabe, also serving as district mayor of surrounding villages	‖‖ Rice paddies and wheat crops lost in Atonoura, Mera, Mikonohama, and Shinjō. Government storehouse flooded in Shinjō.

-han, daimyo domain

Shipping route

Road with paired dots at intervals of 1 ri (4 km)

Castle

Primary sources 根本史料

MIHO

三保

"Miho-mura yōji oboe"

『三保村用事覚』

p. 78-79

EACH ACCOUNT BEGINS at its upper right. The columns read from top to bottom and from right to left (headnote, p. 39). The account reappears with transliteration and translation on the pages identified in italics below the document title.

Each title is enclosed by quotation marks (『 』 in Japanese). Most are names shared by other documents; "Zassho," for instance, means "Miscellaneous records." To make such titles unique we add, outside quotation marks, the name of the family (-ke) or daimyo domain (-han) that produced or preserved the document.

TANABE

田辺

"Tanabe-machi daichō"

『田辺町大帳』

p. 86

JAPAN

Kuwagasaki
Tsugaruishi
Ōtsuchi

Nakaminato
Edo
Miho

Tanabe

N

500 km

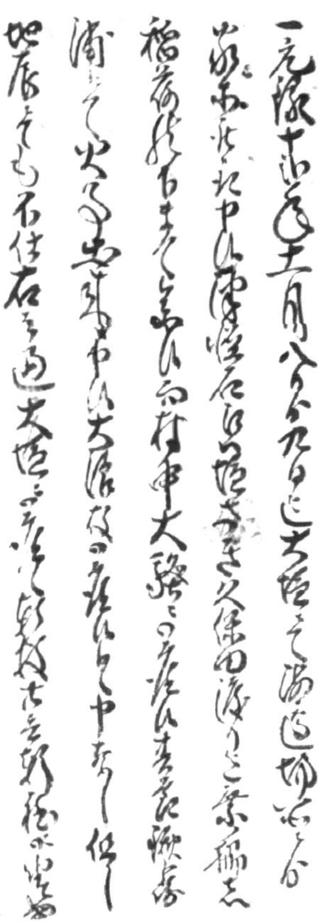

TSUGARUISHI

津軽石

Moriai-ke

"Nikki kakitome chō"

盛合家

『日記書留帳』

p. 52

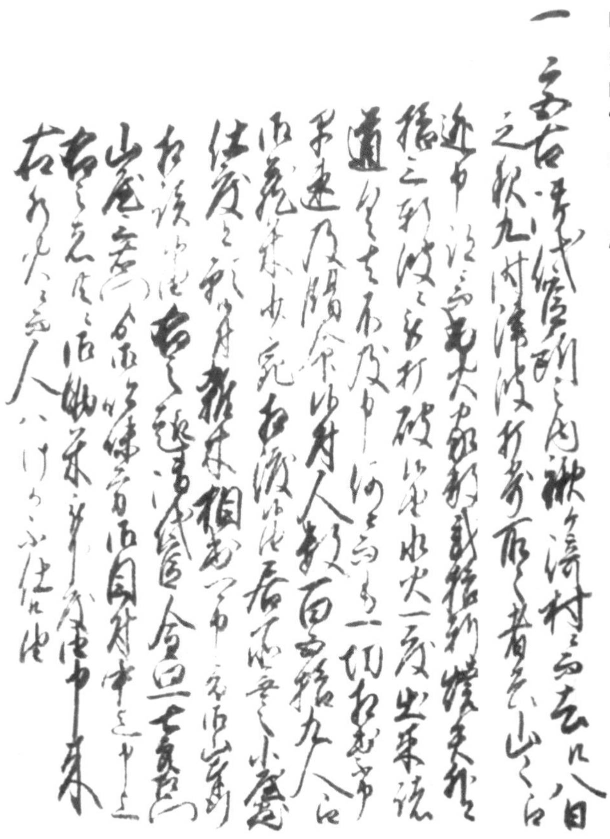

KUWAGASAKI

鍬ヶ崎

Morioka-han

"Zassho"

盛岡藩

『雑書』

p. 38-39

NAKAMINATO

那珂湊

Ōuchi-ke

"Go-yōdome"

大内家

『御用留』

p. 68-69

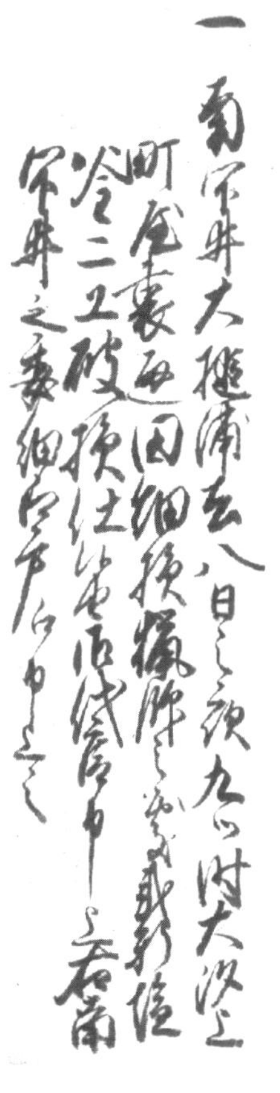

ŌTSUCHI

大槌

Morioka-han

"Zassho"

盛岡藩

『雑書』

p. 60

Kuwagasaki 鍬ヶ崎

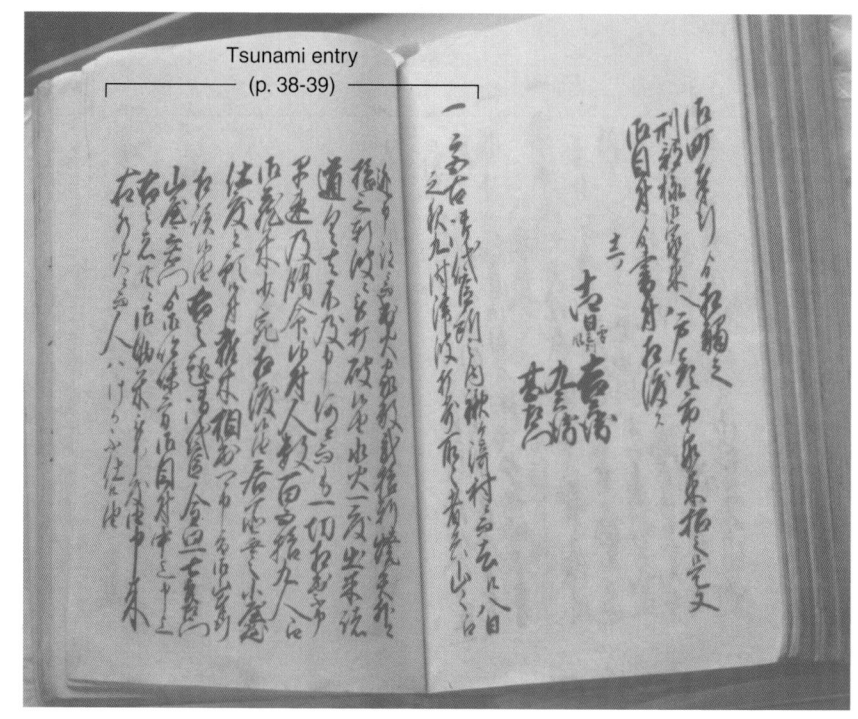

Magistrates' office

MIYAKO

Tax office

Kuwagasaki

Harbor

Miyako Bay

North to Pacific Ocean →

The port of Kuwagasaki was administered from Miyako by district magistrates of a feudal domain, Morioka-han. Administrative records in a volume of Morioka-han "Zassho," compiled by samurai in the domain's castle, mention the 1700 tsunami in Kuwagasaki.

Tsunami entry
(p. 38-39)

KUWAGASAKI had 281 houses a decade or two before 1700 (Takeuchi, 1985a, p. 321, citing Morioka-han "Zassho" for the years 1681-1691). It was then a major port for Morioka-han, as recounted by Iwamoto (1970, p. 116, 119) and implied by a shipping route on the shogunal map from 1702 (dull red line, p. 33).

THE ABOVE VIEW of the village and its surroundings comes from a 1739 map of the Miyako district (p. 44). The tax office arose beside the port in 1701. Its map label reads *jūbun no ichi o-yakuya* ("ten-percent office") because Morioka-han levied a ten-percent tax on non-agricultural goods (Hanley and Yamamura, 1977, p. 129; Iwamoto, 1970, p. 49).

Main points

A nighttime flood and ensuing fires destroyed one tenth of the houses in Kuwagasaki. In response, officials issued food and sought wood for emergency shelters (p. 38-39).

An account of these events, probably written in 1700, calls the flood a "tsunami"—a term used in no other account of the 1700 tsunami in Japan (p. 40-41).

The reported hour of the tsunami in Kuwagasaki, identical to that reported from Ōtsuchi, 30 km to the south, pinpoints the 1700 Cascadia earthquake to the North American evening of January 26, 1700 (p. 42-43).

A regional government run by samurai produced the main account of the 1700 tsunami in Kuwagasaki (p. 44-45).

People went to high ground during the 1700 tsunami, as they did centuries later during the tsunami from Chile in 1960 (p. 46-47).

Waves of the 1700 tsunami directly destroyed 13 houses in Kuwagasaki. The damage in Japan helps define the size of the 1700 earthquake (p. 48-49).

Setting

From the nation's capital in Edo, later renamed Tokyo, the Tokugawa shoguns and their retainers ruled Japan between 1603 and 1867, the Edo period. Under their authority, the Nambu clan controlled much of the northeast part of the nation's main island, Honshu.

The Nambu domain, Morioka-han, included several coastal districts. One of these districts was administered from Miyako. The village of Kuwagasaki, 1 km east of Miyako, adjoined the district's main harbor. The village contained close to 300 houses in 1700.

Other tsunamis

Tsunamis of nearby origin caused deaths in Kuwagasaki in 1611, 1896, and 1933. A lesser near-source tsunami, in 1677, swept away five houses, flooded rice paddies, and damaged salt-evaporation kilns.

Aside from the 1700 event, no tsunami of remote origin is known to have damaged Edo-period Kuwagasaki. The 1960 Chile tsunami entered 14 houses but destroyed none (p. 49).

Documents

Morioka-han "Zassho," an administrative diary compiled in Morioka castle, contains the main account of the 1700 tsunami in Kuwagasaki. The news originated with district magistrates in Miyako. Their report reached Morioka six days after the tsunami (p. 44).

An independent report of the tsunami, dispatched from Ōtsuchi, reached Morioka a day later (p. 60). A merchant's account of the 1700 tsunami in Tsugaruishi mentions, as hearsay, the house fires in Kuwagasaki (p. 52, columns 3-5).

MORIOKA-HAN "ZASSHO," in the volume at left, contains records from the 12th year of the Genroku era (defined p. 42). Each page is 30 cm (12 inches) long; the book weighs 1.26 kg (2.7 lb).

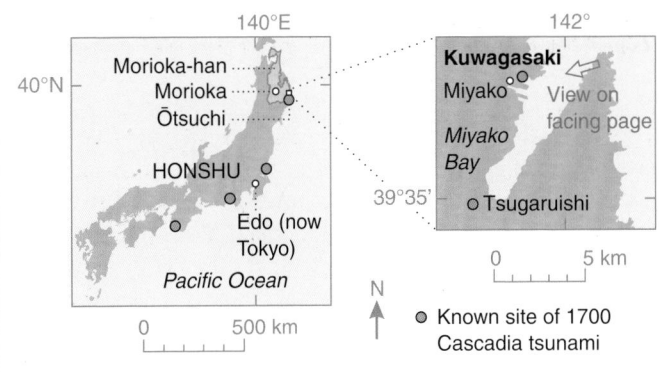

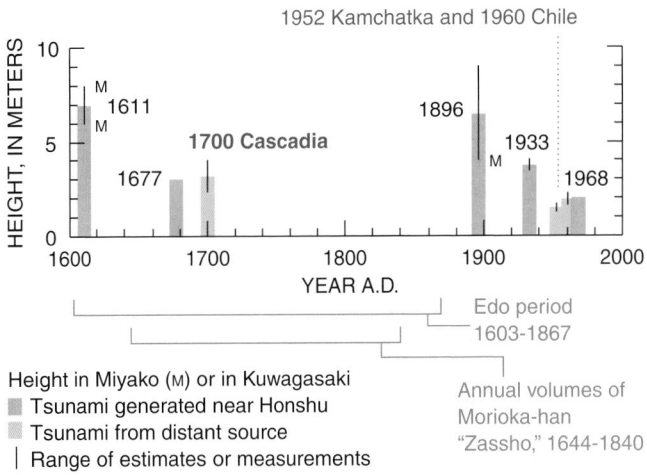

NOTABLE TSUNAMIS IN KUWAGASAKI AND MIYAKO SINCE 1600

Height in Miyako (M) or in Kuwagasaki
- Tsunami generated near Honshu
- Tsunami from distant source
- Range of estimates or measurements

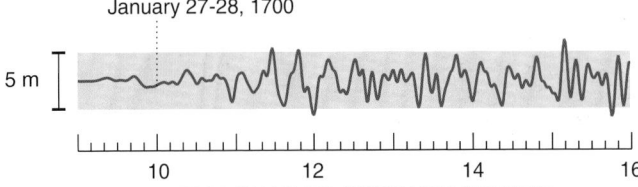

SIMULATED WAVES OF THE 1700 TSUNAMI IN KUWAGASAKI

Tsunami arrives around midnight Japan time, January 27-28, 1700

MEASURED HEIGHTS of the **1952 and later** tsunamis are from The Central Meteorological Observatory (1953, p. 20-22, 46), The Committee for Field Investigation of the Chilean Tsunami of 1960 (1961, p. 178), Unoki and Tsuchiya (1961, p. 258), and Kajiura and others (1968, p. 1370). The 1964 Alaska tsunami, not shown, crested 0.14 m above tide (map, p. 95). The graphed heights of most of the earlier tsunamis were inferred from descriptions of flooding and damage (Hatori, 1995, p. 60; Tsuji and Ueda, 1995, p. 96-97; Usami, 1996, p. 189; pages 48-49 of this report). The **1611** tsunami caused about 100 deaths in Miyako and Kuwagasaki (Hatori, 1995, p. 64; Tsuji and Ueda, 1995, p. 96). In Kuwagasaki alone, the **1896** tsunami killed 125 (Yamashita, 1997, p. 113) and the **1933** tsunami, 24 (Usami, 1996, p. 189).

THE SIMULATED WAVES are those from sea-floor deformation during a magnitude-9 earthquake at the Cascadia subduction zone (p. 98). The wave train lasts more than a day (p. 74-75), like the gauged Chilean tsunami of 1960 (p. 19). The modeled earthquake occurs about 9 p.m. local time on January 26, 1700 (p. 43). Diagram from Satake and others (2003).

Account in Morioka-han "Zassho" 盛岡藩『雑書』の記述

TWELVE CURSIVE COLUMNS in Morioka-han "Zassho" provide an official description of the 1700 tsunami and its aftermath in Kuwagasaki. The tsunami arrived at night (column 2). Villagers fled to high ground (2-3). The water destroyed 13 houses outright (4) and set off a fire that burned 20 more (3). In response, magistrates in nearby Miyako issued rice to 159 persons (6-7) and sought wood for shelters (8-9). They kept others in the han government informed of these emergency efforts (9-12).

The columns contain symbols of Chinese origin (*kanji*) and a few, simpler symbols from Japanese syllabaries (*kana*). The writer applied these symbols with a brush. In gray we

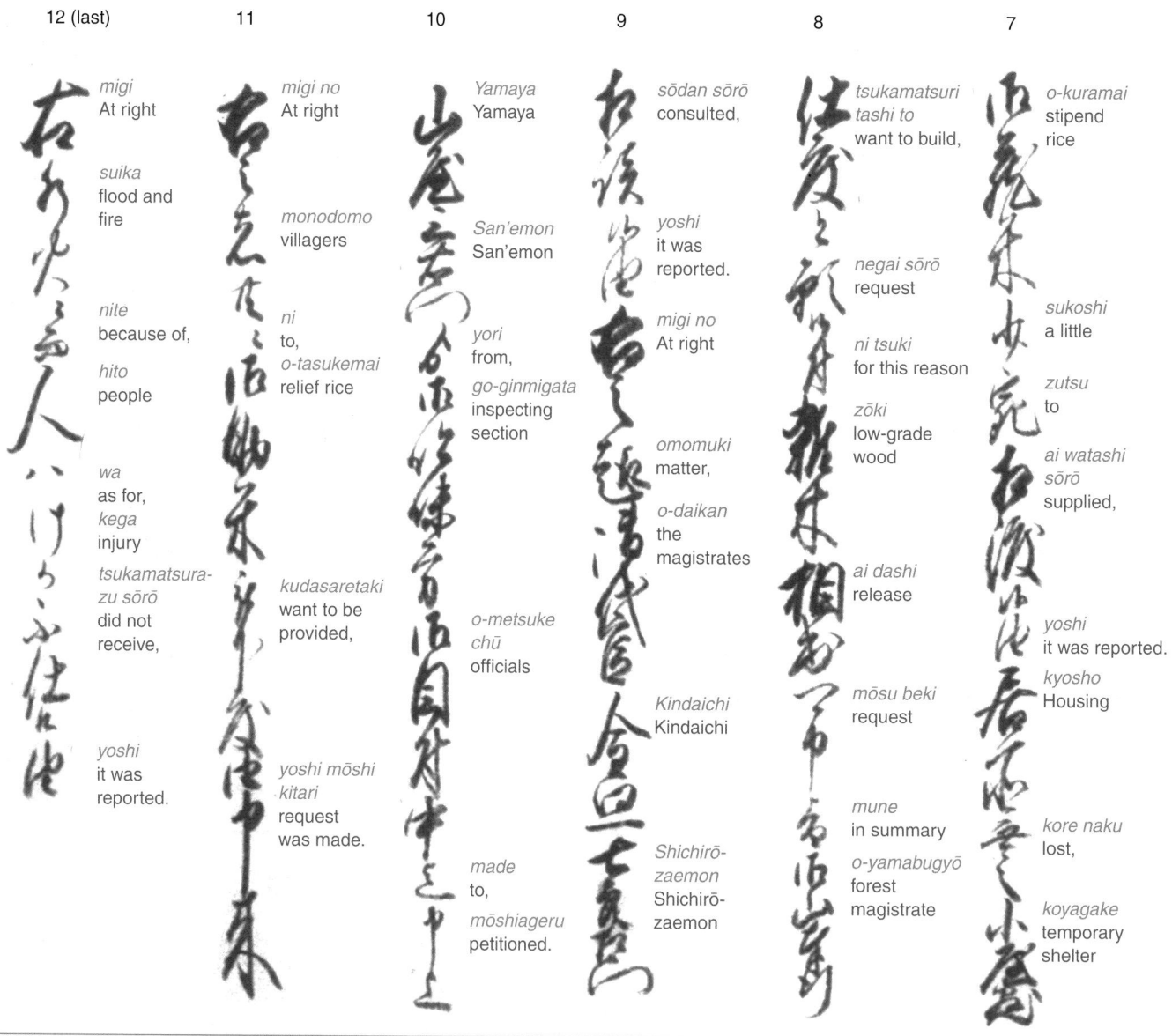

12 (last)	11	10	9	8	7
migi At right	*migi no* At right	*Yamaya* Yamaya	*sōdan sōrō* consulted,	*tsukamatsuri tashi to* want to build,	*o-kuramai* stipend rice
suika flood and fire	*monodomo* villagers	*San'emon* San'emon	*yoshi* it was reported.	*negai sōrō* request	*sukoshi* a little
nite because of,	*ni* to,	*yori* from,	*migi no* At right	*ni tsuki* for this reason	*zutsu* to
hito people	*o-tasukemai* relief rice	*go-ginmigata* inspecting section	*omomuki* matter,	*zōki* low-grade wood	*ai watashi sōrō* supplied,
wa as for, *kega* injury	*kudasaretaki* want to be provided,	*o-metsuke chū* officials	*o-daikan* the magistrates	*ai dashi* release	*yoshi* it was reported.
tsukamatsura-zu sōrō did not receive,			*Kindaichi* Kindaichi	*mōsu beki* request	*kyosho* Housing
yoshi it was reported.	*yoshi mōshi kitari* request was made.	*made* to, *mōshiageru* petitioned.	*Shichirō-zaemon* Shichirō-zaemon	*mune* in summary *o-yamabugyō* forest magistrate	*kore naku* lost, *koyagake* temporary shelter

---

12, *kega tsukamatsura-zu sōrō*—Language reflects the villagers' status below that of the writer.

Formal language—*mōsu* (3, 5, 8), *sōrō* (4, 6-8, 12), *mōsa-zu* (5), *mōshi* (11).

Sound change at word juncture—*doki* for *toki* (2), *domo* for *tomo* in *monodomo* (2), *nijikken* for *nijū-ken* (3), *gen* for *ken* in *jūsan-gen* (4), *issai* for *ichi-sai* (5), *gata* for *kata* in *go-ginmigata* (10).

NOTES, LIKE THE COLUMNS, BEGIN AT RIGHT ON THE FACING PAGE.

yamabugyō commonly worked in the finance office (*kanjōsho*) and reported directly to deputy governors (*karō*) (Totman, 1989, p. 91).

9-10, *Kindaichi...San'emon*—During Genroku 12, the year of the 1700 tsunami (p. 42), four magistrates served in Miyako. Among them were Kindaichi Shichirōzaemon and Yamaya San'emon (Miyako-shi Kyōiku I'inkai, 1991, p. 554).

10, *go-ginmigata*—*go-*, honorific like *o-* in column 1.

8, *tsukamatsuri*—Humble language for addressing a person of higher status. Such deference is shown also by *mōshiageru* (10).

8, *zōki*—*zō*, miscellaneous; *ki*, tree or timber. Probably the writer would have used *mokuzai* had the wood been suitable for fine buildings and furniture.

8, *o-yamabugyō*—Literally, person in charge (*bugyō*) of hills (*yama* as in column 2). In Edo-period domains, senior forest officials called

add Roman letters as a guide to the spoken Japanese (rules, p. v). Literal translations follow in blue.

The columns proceed from right to left. Matter already mentioned therefore appears "at right" (9, 11, 12). Verbs end sentences, some of which are punctuated further by "ink breaths," where bold lines of a newly inked brush start the next sentence (clear example: 右 *migi*, column 9). Nouns follow all their modifiers; prepositions follow their objects.

COLUMN 1 (first)

[start of entry]

Miyako
Miyako

o-daikansho
district magistrate's office

no uchi
within,

Kuwagasaki-mura
Kuwagasaki village

nite
in,

saru
past

yōka
eighth day

2

no
of

yoru
night

kokonotsu-doki
hour of nine,

tsunami
tsunami

uchiyose
came.

shosho no
Here and there,

monodomo
villagers

yamayama
hills

e
to

3

nige mōsu
escaped.

ato nite
Afterwards

shukka
started fire

iekazu
number of houses

niji-kken
20 houses

shōshitsu
burned.

hoka ni
In addition,

4

jūsan-gen
13 houses

nami
waves

ni
by

uchiyaburare sōrō
were destroyed,

yoshi
it was reported.

suika
Flood and fire

ichi-do
at the same time

shuttai
happened.

sho
Various

5

dōgu
belongings

wa
as for,

mōsu ni oyoba zu
needless to say

nani nitemo
everything

issai
at all

ai-dashi
save

mōsa-zu
could not.

6

sassoku
Soon after,

katsumei ni oyobi sōrō
became famished

ni tsuki
thereby

hitokazu
number of people,

hyaku-gojūkyū-nin
159 people

e
to,

6, *katsumei ni oyobi*—*katsu*, thirst; *mei*, life; *ni oyobi*, approach.

7, *o-kuramai*—Rice (*mai*) collected as tax, kept in government storehouses (*o-kura*), and distributed as stipends for samurai. An o-kura adjoined the Miyako magistrates' office in 1692 (Hanasaka, 1974, p. 26-27); the rice in 1700 may have come from this building. The 1700 tsunami entered another o-kura, near Tanabe (p. 86, 88).

2 and 6, *e*—Pronounced and written *e*, means "to."

3, *kken*—The house counter *ken* (like "sheets" in "seven sheets of paper") here follows a slight pause transcribed as a doubled Roman consonant. This same counter changes sound to *gen* in column 4.

5, *ai-dashi*—*ai* adds only emphasis or cadence.

5 and 12, *wa*—Topic marker, written *ha*.

6, *nin*—Counter for people.

← NOTES. Column 1, *Miyako o-daikansho no uchi*—In the district administered from the Miyako magistrates' office (p. 44; office location, p. 36, 49).

1, *o-daikansho*—Honorific *o-* here and in 7-11.

2, *kokonotsu-doki*—Around midnight (p. 43).

2, *monodomo*—Commoners.

2, *yamayama*—More than one hill (*yama*). Second yama denoted by "repeat" symbol,

Words for waves 「津波」を表すことば

In each primary account, the orphan tsunami has a different alias.

SUDDEN SLIP on a submarine fault initiates a typical tsunami while also setting off an earthquake (cartoon, opposite; see also p. 10, 99). Feeling no precursory earthquake (p. 54), several Japanese writers called the 1700 tsunami a high tide. Only the writer of the Kuwagasaki account uses 津波 *tsunami* without questioning the term.

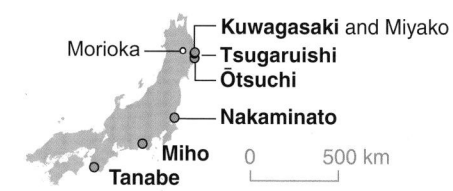

Morioka
Kuwagasaki and Miyako
Tsugaruishi
Ōtsuchi
Nakaminato
Miho
Tanabe

0   500 km

Tsunami 津波

A *tsunami* flooded Kuwagasaki at midnight on January 27-28, 1700, according to the entry in Morioka-han "Zassho" for February 2, 1700. The writer, inland at Morioka (p. 44), was probably paraphrasing a report from coastal magistrates in Miyako. He wrote tsunami with the same pair of symbols used today: 津 *tsu* (harbor) and 波 *nami* (waves).

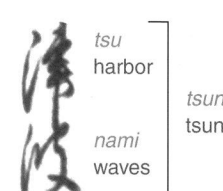

tsu harbor
nami waves

tsunami tsunami

Morioka-han "Zassho," entry dated Genroku 12.12.14 (February 2, 1700) [p. 39, column 2].

High tide 大塩

"High tide" denotes the 1700 tsunami in Tsugaruishi. The symbols mean "big" (大 *ō*) and "salt" or "tide" (塩 *shio*); in context, they connote "high tide." The report states that the flooding was not associated with a felt earthquake.

ōshio high tide

Moriai-ke "Nikki kakitome chō," 18th-century copy. From error in copying, event misdated to Genroku 12.11.8 [p. 52, column 1; 53].

High tide 大汐

In an entry dating to February 3, 1700, Morioka-han "Zassho" uses a different *ōshio* for waters that damaged Ōtsuchi at midnight on January 27-28, 1700. The *ō*, again 大 for "big," modifies 汐 *shio* composed of symbols for water and evening. This same *ōshio* represents the 1700 tsunami in an independent summary of Ōtsuchi magistrates' documents, as later compiled in the printed symbols at right.

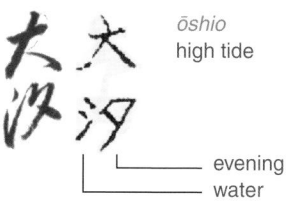

ōshio high tide

evening
water

Left, Morioka-han "Zassho," entry dated Genroku 12.12.15 [p. 60, column 1].
Right, Hand-printed synopsis of "Ōtsuchi kokondaidenki" (Mombushō Shinsai Yobō Hyōgikai, 1943, p. 25) [p. 62, 112].

High waves 浪高久

Nami takaku held a rice boat offshore Nakaminato on January 28, 1700. The waves were probably described as such by the crew; 浪高久 appears in a petition from the captain. At right, the version in a probably 18th-century copy of an accident certificate issued February 12, 1700.

nami waves

takaku high

Ōuchi-ke "Go-yōdome," certificate originally dated Genroku 12.12.24 (p. 70) [p. 69, column 3].

Tsunami? 徒奈三 ?

The headman of Miho village puzzled over what to call the waves in 1700. He described them as "high water" (*mizu takaku*) and as "something like high tides" (*michishio nado no yōni*). He knew and used the term *tsunami* (which he wrote phonetically) but expressed wonder as to why no preceding earthquake was felt in his village or nearby.

tsu

na

mi

tsunami

"Miho-mura yōji oboe," [p. 78, column 10].
The symbols 徒 *tsu*, 奈 *na*, and 三 *mi* are *hentaigana*—variant kana omitted from the simpler, modern syllabaries adopted in the 20th century (46-character hiragana and katakana; Seeley, 2000, p. 143, 153-154).

Unusual seas あびき

A Tanabe municipal record from 1700 introduces the tsunami as あびき *abiki*—unusual seas from tides, storms, winds, or tsunami. The same record also calls the tsunami a "tide" (潮 *shio*). This shio contains a radical for morning, while the one for Ōtsuchi contains "evening"—in accord with times when the tsunami was first noticed (p. 43).

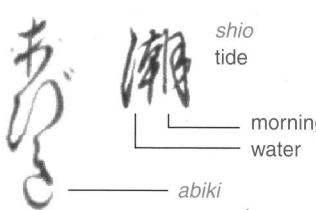

shio tide

morning
water

abiki unusual seas

"Tanabe-machi daichō" [p. 86, columns 1, 3, 5]
The *abiki* can be read as hiragana. On abiki's meanings, see "Nihon kokugo daijiten" (comparable to the Oxford English Dictionary) and a footnote on our page 86. Page 47 tells the story behind the first use of "tsunami" that the OED records.

| 1 | One tectonic plate descends beneath another at a subduction zone. | 2 | Subduction gradually drags the upper plate downward. | 3 | Suddenly released during an earthquake, this plate springs back, making a tsunami. |

 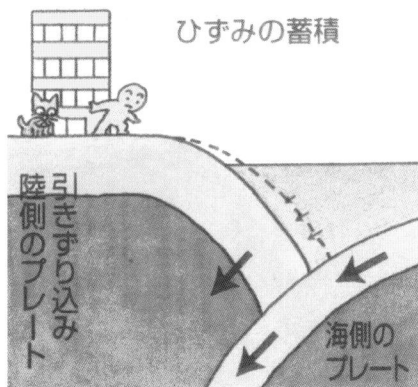 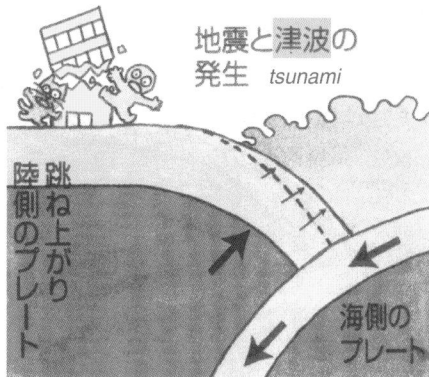

海溝（トラフ）

陸側のプレート　海側のプレート

ひずみの蓄積

引きずり込み　陸側のプレート　海側のプレート

地震と津波の発生 *tsunami*

跳ね上がり　陸側のプレート　海側のプレート

From a public-safety booklet by Jishin Yochi Sōgō Kenkyū Shinkōkai Jishin Chōsa Kenkyū Sentā (undated)

The first "tsunami"

AN OFFICIAL DIARY from 1612 contains what is probably the earliest extant example of 津波 *tsunami*. The writer was an aide to Tokugawa Ieyasu (1542-1616), the first of 15 shoguns to rule Japan from Edo. The document, "Sumpuki," provides a record of Ieyasu's years at Sumpu, near Miho (p. 76), in the last decade of his life.

A tsunami spawned off northeast Honshu took thousands of lives there on December 2, 1611. Among the dead were 100 persons in Miyako, 150 in Tsugaruishi, and 800 in Ōtsuchi. In height the 1611 waves rivaled those of Japan's most disastrous tsunami, which caused 22,000 fatalities in northeast Honshu in 1896.

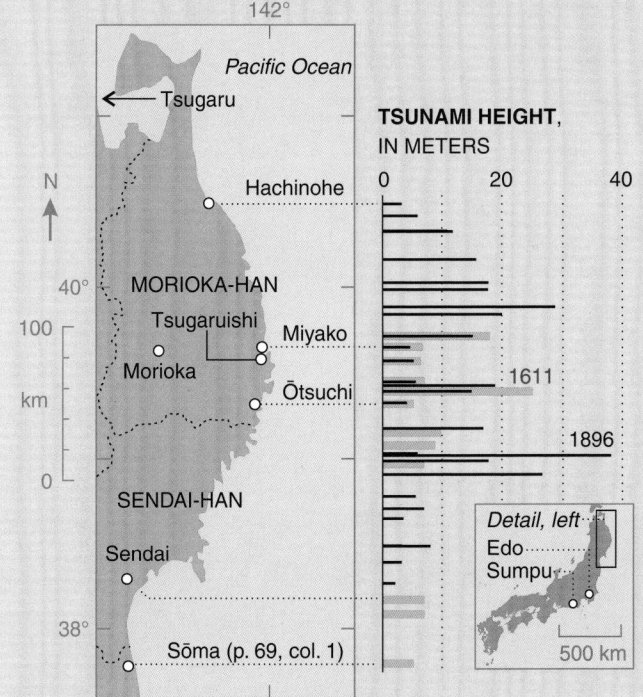

TSUNAMI HEIGHT, IN METERS

Hatori (1995) compiled the heights and 1611 deaths; Yamashita (1997, p. 113), deaths from 1896. For a printed "Sumpuki" see Anonymous (1995).

In Sumpu castle a diarist wrote 津波.

Detail from 1687 map on page 26. East Asian Library, University of California, Berkeley.

The tsunami entry in "Sumpuki," written in January 1612, begins by noting a gift from Date Masamune (1566-1636), daimyo of Sendai-han. From northeast Honshu Masamune has sent *hatsu tara*, the season's first cod.

Aides then tell of the 1611 disaster. A so-called tsunami (*yo ni tsunami to yu'u*) has drowned 5,000 people in the territory of Masamune and 3,000 persons and horses (*jimba*) in the Nambu clan's domain (Morioka-han) and in adjoining Tsugaru. Along with this news comes a story about a samurai who survived the tsunami by faithfully serving his daimyo:

Masamune wants fish. Two samurai receive the order. They round up fishermen. The fishermen balk because the sea has a strange color and the skies look ominous. One of the samurai insists on obeying the daimyo's order. All set out in a boat. Soon it meets the tsunami, which drives it inland into the crown of a pine tree. The waves also sweep away entire villages along the shore. Later, after the water recedes, the men clamber down from the tree. Scanning the shore, they realize that they too would have been swept away had they not gone fishing for Masamune. The two samurai return to Masamune, who bestows a gift upon the one who had insisted on following orders.

The story, as recounted in "Sumpuki," concludes with a moral voiced by Ieyasu: If you follow orders, you may escape disaster and receive gifts.

Converting time 時間の換算

The orphan tsunami came ashore in Japan on January 27 and 28, 1700.

YEARS ON "ZASSHO" COVER

Genroku jūni
Genroku 12

tsuchinoto
younger brother of earth

u
fourth zodiac sign; rabbit

JAPANESE ERAS

Genroku era (1688-1704)

Edo period (1603-1867)

Meiji era

Heisei era

A.D. 1600　　**1700**　　1800　　1900　　2000

TWO CALENDARS date the 1700 tsunami in Morioka-han "Zassho." The year was the twelfth of the Genroku era—one of 35 eras in the period of Tokugawa rule from Edo. Genroku 12 was also a "Year of the Rabbit" in a 60-year zodiacal cycle of Chinese origin.

By either name, the year lasted 384 days; it contained a leap month, between the 9th and 10th month. Each month started on the new moon and lasted 29 or 30 days. The usual 12-month sum came to 354 days—eleven days short of a solar year. Therefore a leap month was inserted every few years, as happened in Genroku 12.

The 1700 tsunami came ashore on the 8th and 9th days of the 12th month of Genroku 12. This final month of Genroku 12 coincides with January and February of 1700. In the Julian calendar, the 1700 tsunami in Japan spans January 17 and 18; in the Gregorian calendar, adopted by Spain in 1582 and England in 1752, the equivalents are January 27 and 28.

CHINESE SIXTY-YEAR CYCLE

JIKKAN	CELESTIAL STEM		*JŪNI-SHI* EARTHLY BRANCH											
	Element	Brother	*ne* mouse	*ushi* bull	*tora* tiger	*u* **rabbit**	*tatsu* dragon	*mi* snake	*uma* horse	*hitsuji* sheep	*saru* monkey	*tori* rooster	*inu* dog	*i* boar
kinoe	wood	elder			51		41		31		21		11 7	
kinoto	wood	younger		2		52		42		32		22		12 8
hinoe	fire	elder	13 9		3		53		43		33		23	
hinoto	fire	younger		14 10		4		54		44		34		24
tsuchinoe	earth	elder	25		15 11		5 1		55		45		35	
tsuchinoto	**earth**	**younger**		26		**16 12**		6 2		56		46		36
kanoe	metal	elder	37		27		17 13		7 3		57		47	
kanoto	metal	younger		38		28		18 14		8 4		58		48
mizunoe	water	elder	49		39		29		19 15		9 5		59	
mizunoto	water	younger		50		40		30		20 16		10 6		60

Year in cycle ⋯⋯　Year in Genroku era

JAPANESE YEARS AND MONTHS

Genroku 12 (leap year; 384 days)

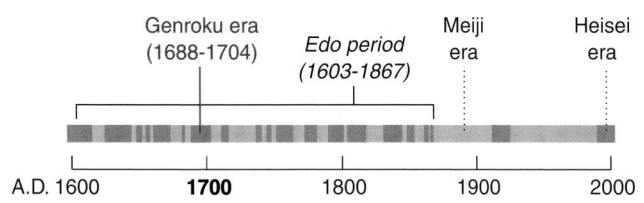

30-day month　29-day month

大 小 大 小 小 大 小 小 大 小 大 大 | 大 大 小 大 小 小 大 小 小 大 小 大 大

1 2 3 4 5 6 7 8 9 u9 10 11 12 | 1 2 3 4 5 6 7 8 9 10 11 12

Rabbit Year waves of Genroku 12, 12th month, 8th and 9th days

Genroku 13 (354 days)

Uru'u kugatsu (leap Ninth Month)

Winter solstice: Genroku 12.11.1 = December 21, 1699 Gregorian

WESTERN YEARS AND MONTHS

J F M A M J J A S O N D | D J F M A M J J A S O N D

1699 Gregorian (365 days)　　**1700 Gregorian** (365 days)

J F M A M J J A S O N D | D J F M A M J J A S O N D

1699 Julian (365 days)　　**1700 Julian** (leap year; 366 days)

Morioka-han "Zassho" records the tsunami's midnight arrival at Kuwagasaki and Ōtsuchi as the hour of nine on the 8th day (p. 39, columns 1 and 2):

kokonotsu-doki
hour of nine

no yoru
of night

yōka
8th day

The writer here refers to traditional Japanese timekeeping that divided a day into in two series of six parts. These "hours," each 120 minutes on average, were counted down, starting at midnight and again at noon, from nine to four:

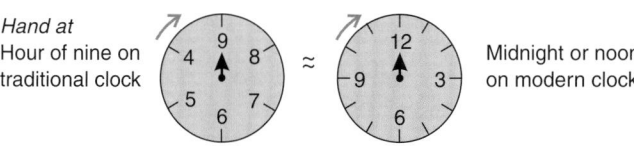

Hand at
Hour of nine on
traditional clock

Midnight or noon
on modern clock

Because the traditional numbered day began at dawn, the 1700 tsunami's midnight arrival in Kuwagasaki and Ōtsuchi refers unambiguously to the 8th day. In Tsugaruishi the flooding reportedly began on the 8th and continued to the 9th. At Nakaminato and Miho, unusual seas were first noticed on the 9th day, in the first hours after dawn. The Tanabe account says "since about dawn of the 8th day"—either near the start of the 8th day or, interpreted for agreement with the other reports, near the start of the 9th.

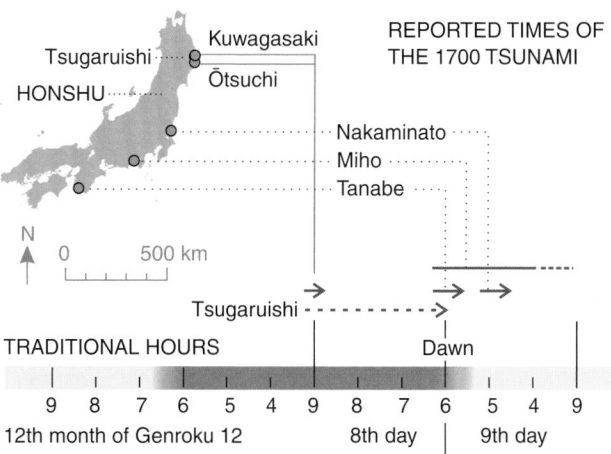

REPORTED TIMES OF
THE 1700 TSUNAMI

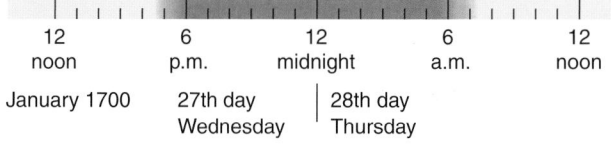

ON CALENDARS AND CLOCKS see Uchida (1975, p. 432), Chamberlain (1905, p. 476-477), Morris (1971, p. 377-382), Nelson and Haig (1997, p. 1251-1256), Parise (1982, p. 294-297), and Steel (2000, p. 2-3).

THE TRADITIONAL HOUR OF SIX coincided with the beginning of dawn and the end of dusk. Its time therefore varied with season, longitude, and latitude. Tsuji and others (1998, p. 9) reckoned that Genroku 12.12.9 dawned in Miho at 6:13 a.m. and in Tanabe ten minutes later.

The 1700 tsunami probably originated in the Cascadia region of North America in response to warping of the seafloor during an earthquake (p. 94, 99). From there, crossing the Pacific Ocean at jetliner speed, the tsunami front needed about ten hours to reach northern Honshu—the arrival simulated on the front cover and on page 75.

Suppose a Cascadia earthquake therefore preceded Kuwagasaki's midnight waves by ten hours. Also allow for a 105-degree difference in longitude (a 17-hour difference in modern time zones). Then the earthquake's local time becomes 9 p.m. Cascadia time on Tuesday, January 26, 1700:

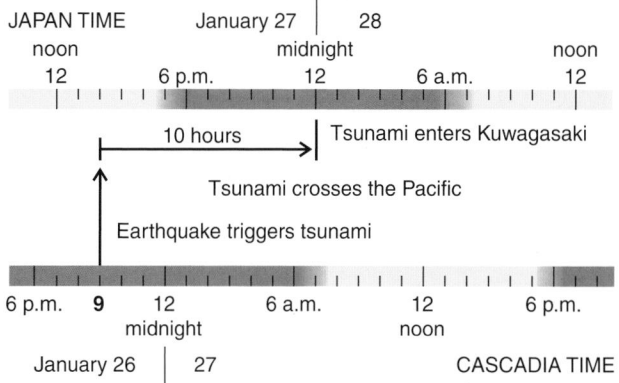

THE TIMELINES above show how Satake and others (1996, 2003) dated the 1700 Cascadia earthquake to about 9 p.m. of January 26. The earthquake may have occurred a few hours earlier if the midnight flooding in Kuwagasaki and Ōtsuchi was from large waves that lagged the tsunami front. Such a lag, between one and two hours long, appears in a simulation of the 1700 tsunami at Kuwagasaki (p. 37) and in Japanese tide-gauge records of the 1960 Chile tsunami (p. 46, 73).

BECAUSE A TSUNAMI'S SPEED in the deep ocean depends solely on ocean depth, a tsunami from Cascadia should take about as much time to reach Japan as does a tsunami that follows a similar path in the reverse direction. Tsunamis from northern Honshu in 1896 and 1933 reached California in 11 hours. Similarly, tsunamis from southern Honshu in 1854 and 1946 reached California in 12-13 hours (box, p. 91; Lander and others, 1993, p. 120, 126, 130, 178).

TWELVE HOURS OR MORE is the likely duration of the 1700 tsunami in Japan. A single train of waves most simply explains the overlap among reported tsunami times (graph at left; p. 72-73). A long train appears in simulations of the 1700 tsunami (snapshots, p. 74-75; graph, p. 37). The 1960 tsunami excited Japanese harbors for days (p. 46, 73).

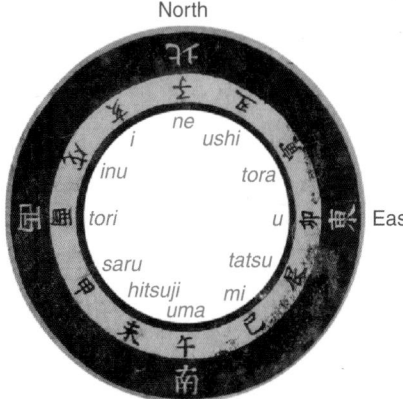

The Chinese zodiac gave the name *u no toshi* (Rabbit Year) that led historians to Miho's full account of the 1700 tsunami (p. 62, 77). Zodiac signs also denoted time of day ("hour of the snake," p. 44) and compass directions (left). The rabbit 卯, as the fourth of twelve signs, faces east 東.

All symbols face outward; 卯 rotates to 辰. From "Nihon kaisan chōriku zu" (p. 30), East Asian Library, University of California, Berkeley.

Samurai scribes 藩の右筆

Military men in a castle town documented their peacetime rule of a feudal domain.

HEADNOTE

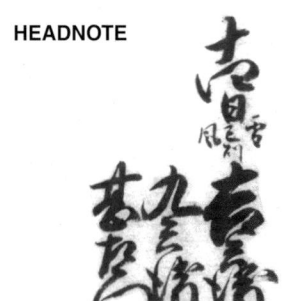

jūyokka
14th day

Yuki minokoku kaze
Snow; wind at the hour of the snake
[about 10 a.m.]

Kichibe'e Kyūbe'e Jinzaemon
Kichibe'e, Kyūbe'e, Jinzaemon
[right]   [left]

AS SNOW FELL on the inland town of Morioka six days after the 1700 tsunami, officials in the administrative wing of Morioka castle received the Miyako magistrates' report about the losses and relief measures in Kuwagasaki. Samurai employed as scribes drafted a synopsis. After review by senior ministers (*karō*, listed above), they entered the final version into the pages of Morioka-han "Zassho" (p. 36).

In similar fashion, officials compiled over 100,000 pages of administrative records in yearly volumes of "Zassho" from 1644 to 1840. One hundred eighty-nine of the yearbooks survive in the air-conditioned attic of Morioka-shi Chūō Kōminkan, a community center.

MORIOKA-HAN "ZASSHO"

Earliest volume, 1644

MORIOKA-HAN "ZASSHO" was kept in Morioka castle until 1874, when the castle was torn down and the "Zassho" volumes entered a storehouse of the Nambu clan. A typeset version has been prepared by Morioka-shi Kyōiku I'inkai and Morioka-shi Chūō Kōminkan (1986-2001). On the han's history and administration see Mori (1972), Hosoi (1988), and Hanley and Yamamura (1977).

THE HEADNOTE dates to six days after the tsunami's midnight arrival in Kuwagasaki on Genroku 12.12.8 (p. 42-43). The year Genroku 12 appears on the volume cover (p. 42); the month is introduced on an interior page.

THE SENIOR MINISTERS, named in full in the year's first "Zassho" entry, were among seven karō who served Nambu Yukinobu (opposite) between 1693 and 1702 (Hoshikawa and Maezawa, 1984-1985; Yoshida and Oikawa, 1983-1992). Nakano Hiroyasu (ca. 1658-1745) served as Kichibe'e in 1690-1713. He inherited the post as an adopted son of the preceding karō Kichibe'e. Kita Yoshitsugu(?) (1670-1732) succeeded his own father as Kyūbe'e in 1696. He held that post for seven years and later returned to it, after the death of his son, for another six. Urushido Shigetada (1640-1709), born in Edo, was Jinzaemon in 1691-1700.

"ZASSHO" EARTHQUAKES

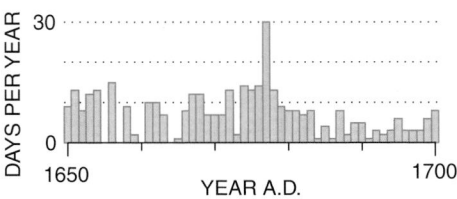

Morioka-han "Zassho" abounds in incidental facts. Headnotes summarize daily weather, which we use to confirm an error in the orphan tsunami's dating (p. 52-53) and to rule out storms as the cause of flooding (p. 72). Graphed above are the yearly number of days when "Zassho" scribes noted an earthquake or earthquakes. The years 1650-1700 included 366 such earthquake days.

Official papers of Morioka-han also include large picture maps (*ōezu*). The one being unfolded below, from 1739, shows one of the domain's 33 administrative districts. The maps aid in envisioning the bygone villages and fields that were flooded in 1700 (p. 36, 50, 56, 58).

DISTRICTS
OF MORIOKA-HAN

Nambu lands of Hachinohe-han (p. 52, footnote) transferred from Morioka-han in 1664

Morioka-machi

○ Miyako

○ Ōtsuchi

Scale bar 240 km = length of Pacific coast of Washington state (p. 8)

N
↑

PICTURE MAP
OF MIYAKO DISTRICT, 1739

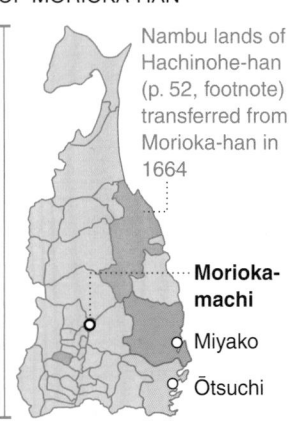

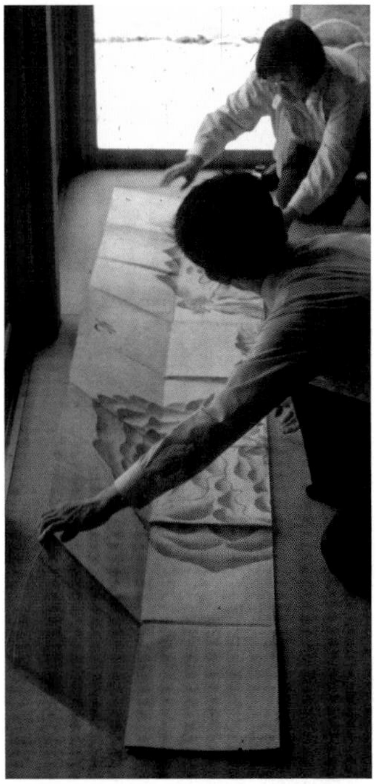

MORIOKA
-han, feudal domain
-machi, castle town
-shi, modern city

EARTHQUAKES recorded in Morioka-han "Zassho" were tabulated by the Ōfunato City Museum in a well-attributed list of earthquakes and tsunamis in northeast Honshu (Ōfunato Shiritsu Hakubutsukan, 1990). The graph above was compiled from pages 9-29 of that list, in the manner described by Satake (2002). The peak of 30 earthquake days in 1677 results from aftershocks of the earthquake whose tsunami washed away five houses in Kuwagasaki and flooded the main street in Ōtsuchi (p. 37, 59).

THE MAP OF HAN DISTRICTS (*tōri*; with sound change at word juncture, *dōri*) is from Morioka-shi Chūō Kōminkan (1998, p. 26). The ōezu of Miyako-dōri, conserved at the Kōminkan, dates to Genbun 4 (1739).

The scribes of Morioka-han "Zassho" resided in Morioka-machi, a castle town zoned by class. As samurai, they probably lived nearer the castle than did artisans and merchants—commoners whose neighborhoods ringed the town. A census in 1700 identified 14,209 commoners in Morioka-machi, out of a total of 343,499 in Morioka-han.

The scribes worked in the *ninomaru*, an administrative center that adjoined the keep of Morioka castle. They may have trained at a military school just north of the castle grounds.

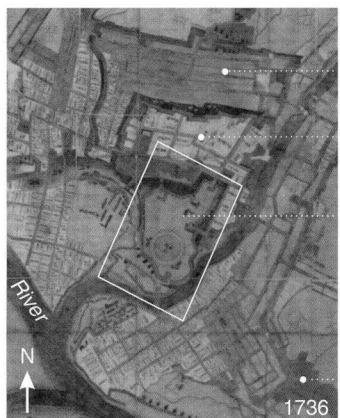

CASTLE TOWN

Commoners' neighborhood
Merchants and artisans

Samurai neighborhood
Individuals' names are plotted.

Morioka castle White box, area of detail below

1736

Temple or shrine

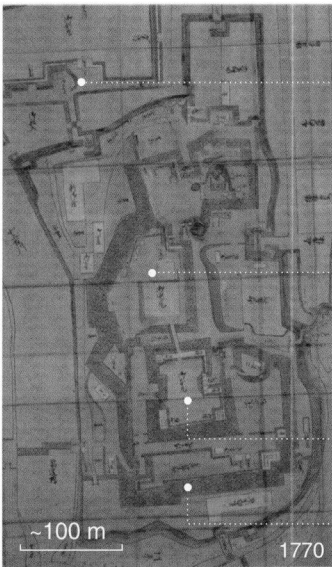

CASTLE GROUNDS

School (site) Founded by 1672, moved elsewhere in 1740

Ninomaru Administrative center where Morioka-han "Zassho" was written and stored

Honmaru Castle keep

Castle wall

~100 m

1770

A monument stone in a city park now marks the site of the *ninomaru* of Morioka castle, demolished in 1874.

2002

IN CASTLE TOWNS of Edo-period Japan, samurai typically resided nearest the castle (Sakudō, 1990, p. 149). The census figures come from Mori (1963, p. 641-643, 688-692); the 1736 map, from Morioka-shi Chūō Kōminkan (1998, p. 2); the 1770 map, from a visitors pamphlet for Morioka-shi Chūō Kōminkan. Nagaoka (1986, p, 81-82) locates the school, *Han go-shinmaru go-keikojo*, which probably began as a military institute for the sons of high-ranking samurai.

DAIMYOS

Nambu Nobunao Subdued opposition in the 1590s

Nambu Yukinobu Ruled in peacetime 1693-1702

Morioka-han "Zassho" written 1644-1840

Morioka castle built 1598-1633

Edo period

A.D. 1500   1700   2000

Scribes in Morioka castle and district magistrates in Miyako served the Nambu clan. The Nambu wrested control over the region while Japan was emerging from centuries of civil war. Nambu rule then continued for more than 250 years under Tokugawa shoguns, throughout the Edo period.

This Nambu authority initially depended on Toyotomi Hideyoshi, who gained hegemony over Japan's warlords in the 1580s and 1590s. In July 1590, Hideyoshi granted much of northeast Honshu to Nambu Nobunao on condition that he survey taxable land and subjugate its owners. Nobunao faced a farmer's rebellion a few months later and an attempted coup the following March. Hideyoshi sent an army to his aid in August 1591.

Morioka castle, begun under Nobunao in 1598, became the domain's headquarters in the 1630s. By then, power had passed from Hideyoshi to the Tokugawa shogunate, and Nambu retainers were attending to peacetime administration. Their successors, under the daimyo Nambu Yukinobu, documented the 1700 tsunami.

Nambu family crest
A pair of cranes symbolizes the clan that controlled Morioka-han throughout the Edo period.

PORTRAITS AND CREST courtesy of Morioka-shi Chūō Kōminkan. Totman (1993, chap. 4) outlines pacification under Hideyoshi and Ieyasu. Han were reorganized into today's prefectures after 1871 (Beasley, 1982, p. 126).

High ground 高台へ

In 1700, Japanese villagers used a timeless tactic for surviving tsunamis.

1700 tsunami

A curling wave chases a fleeing figure on roadside signs in Washington, Oregon, and California. In Sumatra in 2004, how many lives would such signs have saved?

In Japan in 1700, tsunami-savvy villagers fled to high ground during a tsunami that originated not far from where the North America signs stand. Going to hills probably prevented casualties at Kuwagaski; all its villagers escaped injury despite a sea flood that destroyed 13 houses and set off a fire that burned 20 more. Miho's headman advised the elderly and children to go to a shrine on the area's highest ground.

THE HIGHWAY SIGN, designed in Oregon in the late 1990s, was adopted as an international symbol in 2003 (http://ioc3.unesco.org/itic/contents.php?id=71). QUOTES are from Morioka-han "Zassho" (p. 38-39, columns 2-3) and "Miho-mura yōji oboe" (p. 79, columns 6-7).

KUWAGASAKI

monodomo
villagers

yamayama
hills

e
to

nige mōsu
escaped

MIHO

mura no
village's

rōnyaku
old and young

o-miya
shrine

e
to

nigashi
[I advised them to] escape

1960 tsunami

As a precaution against the incoming tsunami from Chile, local police and firemen began evacuating low-lying parts of Kuwagasaki at 4:05 a.m. on May 24, 1960. This action anticipated, by an hour, a tsunami warning from the regional headquarters of the Japan Meteorological Agency, in Sendai. The local officials used tips like these, which weather observers received at JMA's station in Kuwagasaki:

3:30....Chilean tsunami in Hawaii (*source:* radio news)
3:40....Surge at Norinowaki (fishermen's association)
3:49....High water in Ōtsuchi (town officials)
3:55....High water near Kuwagasaki (local resident)

Soon after 4:13 sunrise, with the local evacuation underway, JMA observers in Kuwagasaki saw the sea withdraw. By then, they had received a further inquiry from the fishermen's association and official calls from Miyako and Yamada. By 4:30, the sea had flooded a pier in Kuwagasaki and a wave had been reported from Tarō.

Forerunner waves prompted the 3:40 report from Norinowaki. Such advance notice of the 1960 Chile tsunami registered on 27 Japanese tide gauges. Did forerunners similarly warn of the 1700 Cascadia tsunami in Japan?

MIYAKO WEATHER STATION in 1960 stood on low ground in Kuwagasaki (map, p. 49). The station report is in Japan Meteorological Agency (JMA) (1961, p. 232-233). JMA Sendai had authority to issue a warning; JMA Miyako did not.
THE MARIGRAM was recorded by the Miyako tide gauge, Kuwagasaki (JMA, 1961, p. 271; map, p. 49). There, 1960 mean sea level stood 0.1 m above a national vertical datum, TP (Tokyo Peil). Though the marigram shows the tsunami cresting 1.2 m above TP (JMA, 1961, p. 34), a high-water line 2.4 m above TP was observed on a warehouse near the gauge (p. 49). For details on forerunner waves of the 1960 tsunami, see Nakamura and Watanabe (1961); on damping of tsunamis by tide gauges, Satake and others (1988).

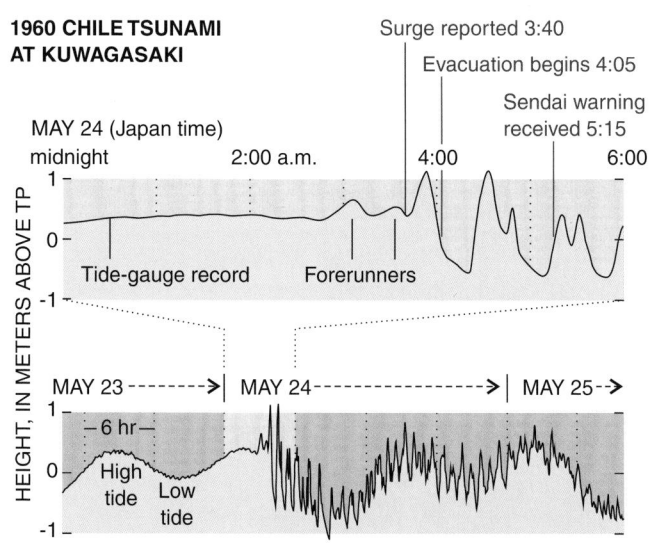

The tsunami began off Chile nearly 24 hours before reaching Japan.

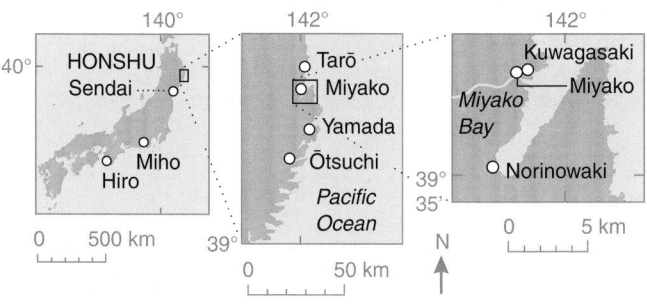

1854 tsunami

Hamaguchi Gohei, a village elder, knows "all the traditions of the coast." One autumn evening, high above the seaside village that he heads, the old man feels an earthquake "not strong enough to frighten anybody." Soon the sea withdraws in a "monstrous ebb." As unknowing villagers flock to the beach, Gohei torches his rice harvest—"most of his invested capital." The villagers rush uphill to fight the fire. Their headman's selfless ruse has saved them from a tsunami.

As "Inamura no hi" ("The rice-sheaf fire"), this story first appeared in a Japanese grade-school reader in 1937 (p. 113) and later appeared in video (p. 115-121). As "The Wave," an American children's book, the tale similarly became a video sent to hundreds of schools in the 1990s for tsunami education in British Columbia, Washington, Oregon, and California. As public art inscribed in stone in Seattle, a "true story" tells of "an old farmer in Japan who saved an entire village from destruction by a tidal wave."

The story, timeless as a cautionary tale about natural tsunami warnings, originated in 1897 as "A living god" by Lafcadio Hearn. Hearn blended two 19th-century disasters: Honshu's giant waves of 1896 (p. 41), whose parent earthquake was weak; and a tsunami evacuation a half century earlier in the southwest Honshu village of Hiro. On the night of December 24, 1854, 34-year-old Hamaguchi Goryō lit rice-straw fires in Hiro during a tsunami that shortly followed a violent earthquake of estimated magnitude 8.5. Lost that night in Hiro were 36 lives and 158 of 374 houses. Goryō himself nearly drowned.

Hamaguchi Goryō's rice-straw fires in Hiro village beckoned villagers well-rehearsed in seeking high ground. As a precaution against tsunamis, Hiro had already evacuated twice in 1854—after earthquakes on July 9 and December 23. Goryō's beacons on December 24, depicted below, are said to have guided nine persons to safety.

HIGH GROUND

Rice-straw fire   Line of fleeing villagers

Paddies

TSUNAMI

HIRO VILLAGE

Yuasa village, 1 km

SEA

HAMAGUCHI GORYŌ, who founded Hiro's first public school in 1852, underwrote the village's recovery from the 1854 disaster (Tsumura, 1991; Shimizu, 1996). Drawing on proceeds from his family's soy-sauce business (today's Yamasa brand), he employed villagers to rebuild 50 houses and to replace the seawalls being overwashed in the painting, above. His wall, still standing, rises 5 m and extends 0.6 km. Hiro village is now the town of Hirogawa.

THE PAINTING above, courtesy of Yōgen temple in Hirogawa, portrays Goryō's fires as numerous and widely dispersed in rice paddies being covered by the tsunami of December 24, 1854. The artist, Furuta Shōemon (also known as Furuta Eisho), witnessed the tsunami in adjoining Yuasa village (Shimizu, 2003).

"A LIVING GOD," quoted in the first paragraph above, contains the first use of *tsunami* cited in the Oxford English Dictionary: "'*Tsunami*!' shrieked the people; and then all shrieks and all sounds and all power to hear sounds were annihilated by a nameless shock heavier than any thunder, as the colossal swell smote the shore with a weight that sent a shudder through the hills..." (Hearn, 1897, p. 7).

"INAMURA NO HI" remained, until 1947, in a national reader for students 10-11 years old (p. 113). Nakai Tsunezō, a grade-school teacher, wrote this adaptation. Hodges (1964) similarly made "A living god" into "The wave." Ellen Ziegler, with a 1987 grant from the Seattle Arts Commission, inscribed the tale on tablets fronting Jefferson Park fire station.

Tsunami size 津波の高さ

Kuwagasaki's losses in 1700 suggest a tsunami several meters high.

WRITTEN RECORDS of flooding and damage by the 1700 tsunami imply tsunami heights in the range 2-5 m in Japan. The heights, graphed in blue below, are central to the conclusion that the Cascadia subduction zone can produce earthquakes of magnitude 9 (p. 98-99).

The height estimates vary with assumptions about the recorded damage and flooding, illustrated for Kuwagasaki in the cartoons at right. In an estimate that is probably low (**A**), the 1700 tsunami crested 2½ m high when it destroyed 13 houses in Kuwagasaki. The estimate increases to 3 m under moderate assumptions (**B**), and it reaches 4 m if it includes a generous assumption about land-level changes since 1700 (**C**; p. 65).

Summary of tsunami heights, 1700 and 1960

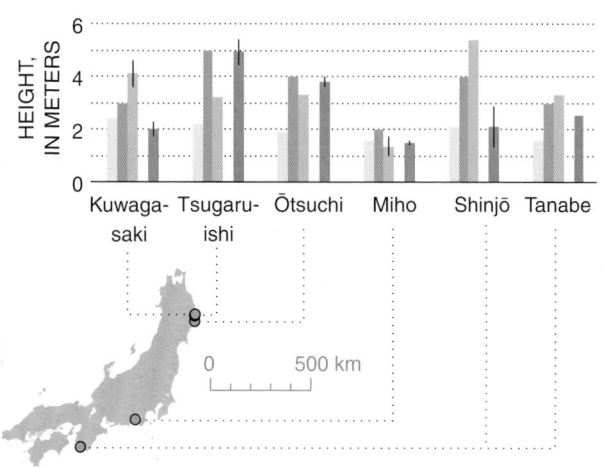

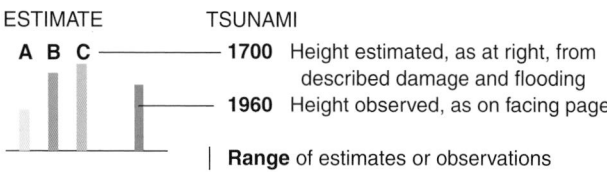

ON TSUNAMI HEIGHTS in 1700, see also pages 56-57, 64-65, 82-83, and 88-91. **A** and **B**, "low" and "medium" height estimates of Satake and others (2003); **C**, estimate by Tsuji and others (1998). TP, datum near sea level (p. 46, footnote).

AT THE MIYAKO GAUGE, high tide averaged 0.6 m above mean sea level for the years 1954-1959; the highest tide in 1937-1995 was 1.0 m above 1995 mean sea level (Japan Meteorological Agency, 1960, p. 402; 1996, p. 253).

WHERE THEY DESTROYED wooden Edo-period houses in Japan, tsunamis commonly ran at least 1.0-1.5 m deep (Tsuji, 1987).

Estimates of 1700 tsunami height in Kuwagasaki

REPORTED DAMAGE

jūsan-gen
13 houses

nami ni
waves by

uchiyaburare sōrō
were destroyed

Morioka-han "Zassho" (p. 39, column 4)

HEIGHT ESTIMATES, IN METERS ABOVE TIDE

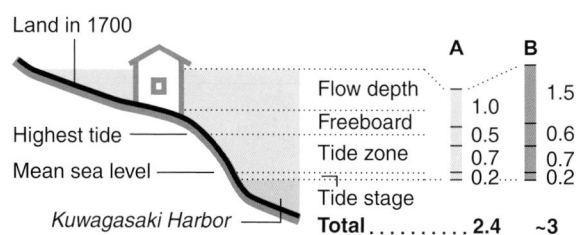

	A	B
Flow depth	1.0	1.5
Freeboard	0.5	0.6
Tide zone	0.7	0.7
Tide stage	0.2	0.2
Total	2.4	~3

ASSUMPTIONS

Flow depth Tsunami flowed 1.0-1.5 m deep where it destroyed houses (1.0 m assumed in **A**, 1.5 m in **B**, 1.0-1.5 m in **C**).

Freeboard To escape storm waves—and perhaps with the 1677 tsunami in mind (p. 37)—villagers built their houses at least 0.5 m above highest astronomical tide (0.5 m in **A**, 0.6 m in **B**).

Tide zone The highest astronomical tide in 1700 was 0.7 m above mean sea level, by analogy with modern tides in Kuwagasaki Harbor (Miyako gauge; location on facing page). Used in **A** and **B**.

Tide stage Tide stood 0.2 m below mean sea level when houses were destroyed (**A**, **B**, and **C**; supporting details, p. 83).

Modern ground The ground surface, surveyed in 1971-1995, was 1.4-1.9 m above TP in probable area of 1700 destruction (**C**).

Subsidence Relative to the sea, land at Kuwagasaki has subsided about 1 m since 1700 (**C**; p. 65).

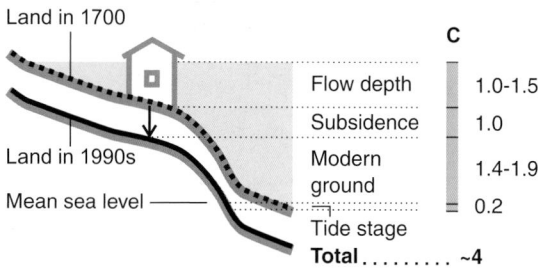

	C
Flow depth	1.0-1.5
Subsidence	1.0
Modern ground	1.4-1.9
Tide stage	0.2
Total	~4

1960 tsunami

In Japan, the 1700 tsunami probably attained about the same size as the 1960 tsunami (summary graph, opposite)—the trans-Pacific waves that began in Chile (p. 10-11, 19).

In Kuwagasaki, where the 1700 tsunami probably reached heights of 2-3 m, the 1960 tsunami crested about 2 m high. The range of 1960 heights is 1.8-2.3 m above a datum near mean sea level (map, below). This range corresponds to 1.7-2.2 m above tide because the 1960 tsunami crested when the tide was about 0.1 m above this datum (p. 46, tide-gauge record). The 1960 tsunami in Kuwagasaki wetted floors of 14 houses without destroying any buildings. In the photo at right, shallow flooding by the tsunami has littered a street with empty petroleum barrels.

EFFECTS OF SHALLOW FLOODING IN 1960

Probable high-water line

VIEW northward from tip of arrow on map below. At left, in a grocery, the 1960 tsunami crested about 0.3 m above a street-level floor (Atwater and others, 1999, p. 9). Photo from Japan Meteorological Agency (1961, p. 351).

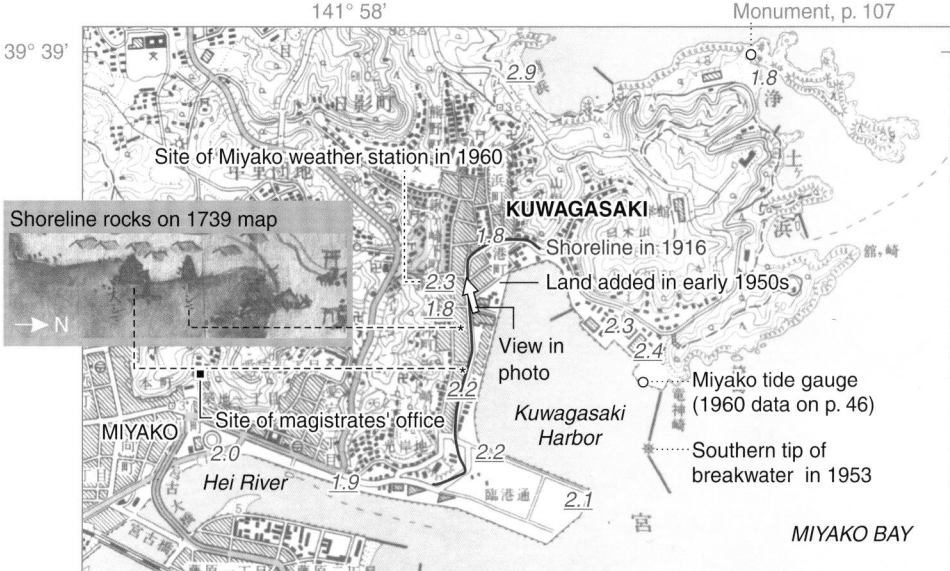

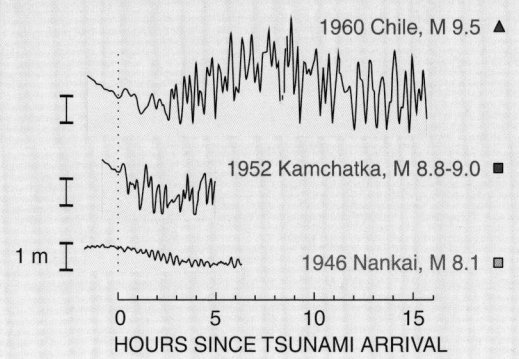

Shoreline rocks on 1739 map

Height of 1960 tsunami (m above TP)
_2.0_ Value with "highest accuracy"
1.8 Value with "moderate accuracy"

N
Scale 1:25,000
0   1 km

TSUNAMI HEIGHTS from The Committee for Field Investigation of the Chilean Tsunami of 1960 (1961, p. 178). Height "2.3" denotes point 4 of Unoki and Tsuchiya (1961, p. 258).

FORMER SHORELINE plotted from Miyako 1:50,000-scale topographic map published in 1917 by Rikuchi Sokuryōbu and updated in 1953 by Naimushō Chirichōsasho. Shoreline rocks on a map at the Miyako fire department.

MAGISTRATES' OFFICE location from Hanasaka (1974, p. 26-27).

BASE MAP from Kokudo Chiriin (Geographical Survey Institute), Miyako 1:25,000 quadrangle, 1996. Contour interval 10 m.

Tsunami height and earthquake size

IN RELATING TSUNAMI DAMAGE in Japan to earthquake size at Cascadia, we assume that the height of a far-traveled tsunami increases with the size of the parent earthquake. The graphs at right show this tendency among 14 subduction-zone earthquakes of magnitude 8.1-9.5 between 1933 and 1974 (p. 98). Below, a similar trend among tide-gauged tsunamis at Crescent City, California.

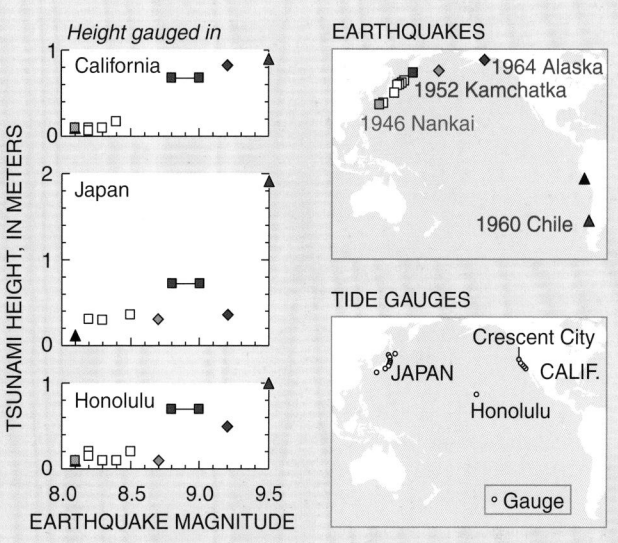

1960 Chile, M 9.5 ▲

1952 Kamchatka, M 8.8-9.0 ■

1 m

1946 Nankai, M 8.1 □

0   5   10   15
HOURS SINCE TSUNAMI ARRIVAL

Height gauged in
California
Honolulu
Japan

TSUNAMI HEIGHT, IN METERS

EARTHQUAKES
1964 Alaska
1952 Kamchatka
1946 Nankai
1960 Chile ▲

TIDE GAUGES
Crescent City
JAPAN   CALIF.
Honolulu
○ Gauge

8.0   8.5   9.0   9.5
EARTHQUAKE MAGNITUDE

MARIGRAMS from Lander and others (1993, p. 178, 181, 193). SCATTER PLOTS show maximum heights among gauges at least 1,000 km from the tsunami source (Abe, 1979, p. 1562). On size of the 1952 Kamchatka earthquake, see Johnson and Satake (1999).

Tsugaruishi 津軽石

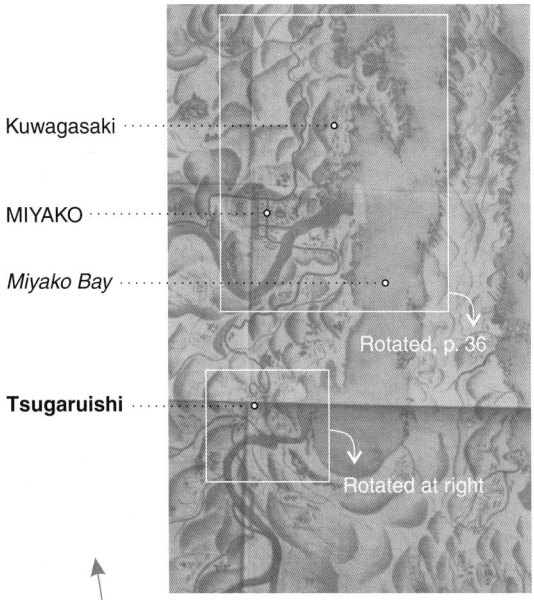

Kuwagasaki

MIYAKO

Miyako Bay

Tsugaruishi

Rotated, p. 36

Rotated at right

Typical north in left map;
orientation and scale vary

Approximate north
in detail at right

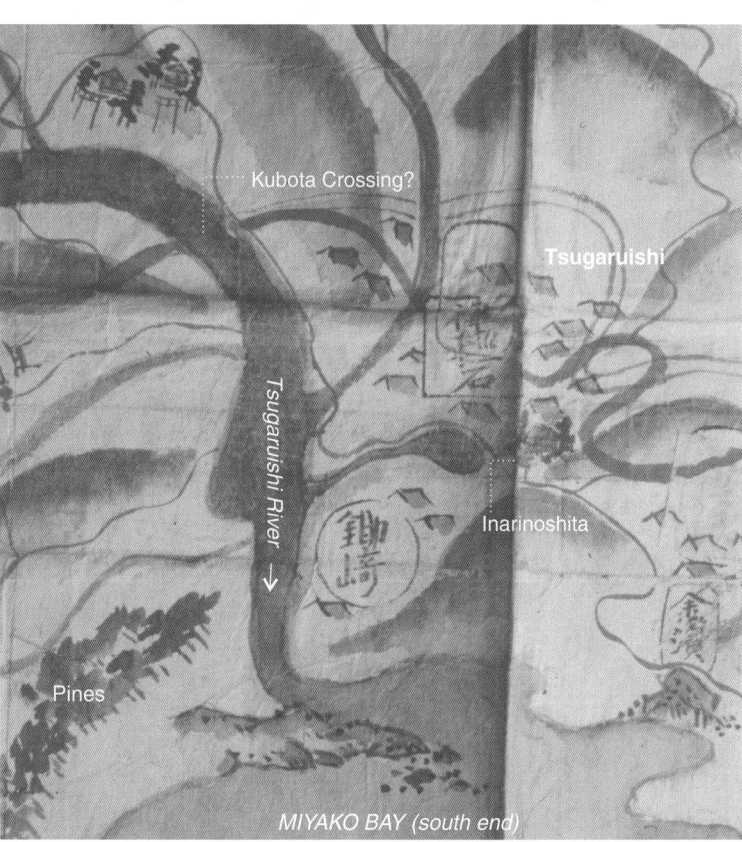

Kubota Crossing?

Tsugaruishi

Tsugaruishi River

Inarinoshita

Pines

MIYAKO BAY (south end)

The south end of Miyako Bay, 7 km from Kuwagasaki,
flanked Edo-period villages near the mouth of a river
known for its salmon. The largest village, Tsugaruishi,
adjoined the river 1 km upstream from the bay.

2004

Moriai-ke "Nikki kakitome chō," a Tsugaruishi
family's notebook for the years 1696-1703, mentions
the 1700 tsunami as high water that swept away
houses along the bayshore, went inland to
Inarinoshita and Kubota Crossing, and reportedly
caused a related fire in Kuwagasaki. "Nikki" further
states that there was no accompanying earthquake.

UPPER VIEWS from the 1739 map of Miyako-dōri (unfolding,
p. 44). Scale varies from place to place on the map, as does
perspective shown by brown rooftops and red shrine gates
(additional examples, p. 36, 56).

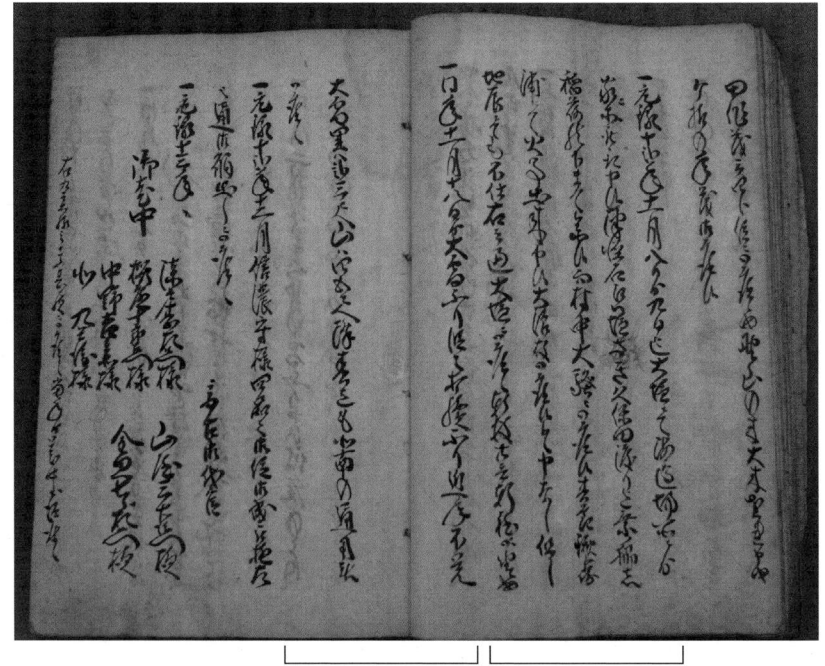

Entries about   Heavy snow   Tsunami (p. 52)

Both events misdated by exactly one month
(p. 53)

Main points

High water at the south end of Miyako Bay washed away houses and entered Tsugaruishi village, 1 km inland. The same event set off a fire that burned "about 21 houses" in Kuwagasaki (p. 52; compare p. 39, col. 3).

The flooding happened without an earthquake (p. 54).

The water went upvalley to "Kubota Crossing"—perhaps as far as did the 1960 Chile tsunami, which ran 2 km inland from the south shore of Miyako Bay. Therefore the 1700 tsunami may have attained heights like those of the 1960 tsunami—about 5 m at the bayshore (p. 56-57).

The Tsugaruishi account originated with a merchant family that built a local financial empire in the 18th century (p. 53).

Setting

Tsugaruishi village, today as in 1700, occupies alluvial fans 1 km south of Miyako Bay. East of the village a farmed plain extends northward to a pine-covered beach ridge near Akamae. Pines also bordered this part of Miyako Bay in 1739 (detail, opposite; mapped also on p. 56).

Edo-period Tsugaruishi belonged to the Miyako district of Morioka-han (p. 44). In the 1680s the village contained 183 houses—about 100 fewer than Kuwagasaki.

Other tsunamis

Tsunamis from earthquakes along the coast of northeast Honshu took lives at the south end of Miyako Bay in 1611, 1896, and 1933. A lesser near-source tsunami in 1677 swept away 13 houses in Kanahama and ten houses in Akamae while damaging 70 hectares of rice paddies near Tsugaruishi.

The 1960 Chile tsunami resonated in Miyako Bay. Just 2 m high along the Pacific coast, the waves rose inside the bay and crested about 5 m high at its south end (p. 55). From there the waters ran past Norinowaki and Tsugaruishi to a limit 2 km inland (p. 56).

Documents

Earthquake researchers learned of the 1700 tsunami in Tsugaruishi through a 1983 transcription by a noted regional historian, Mori Kahei. In 1993 they quoted this transcription in the earthquake anthology "Shinshū Nihon jishin shiryō" (p. 62, 123).

In 2004 we viewed Mori's source document in the home of the Moriai family of Tsugaruishi (home interior, p. 53). That source is a Moriai family notebook for the years 1696-1703 (opposite). Because of a copyist's error, the notebook dates both the orphan tsunami and a subsequent snowstorm exactly one month early (p. 52-53)

THE 17TH-CENTURY STATISTICS on Tsugaruishi can be found in Iwamoto (1970, p. 11) and Takeuchi (1985a, p. 507).

MORI KAHEI reviewed accounts of tsunamis of northeasternmost Honshu (Mori, 1983, p. 155-175). His transcription of the "Nikki" account (p. 161) contains a small error in transcribing "Norinowaki" (footnote, p. 52). For the quote in "Shinshū," see Tokyo Daigaku Jishin Kenkyūsho (1993, p. 146).

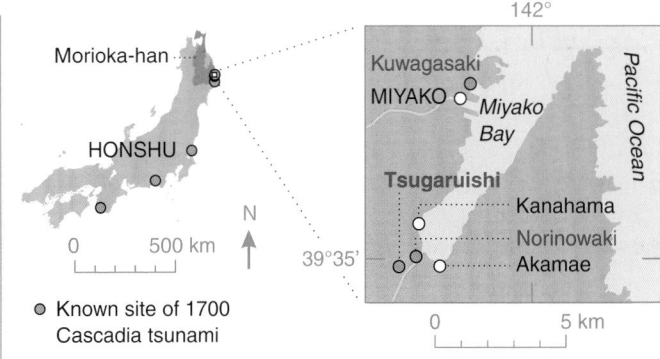

Known site of 1700 Cascadia tsunami

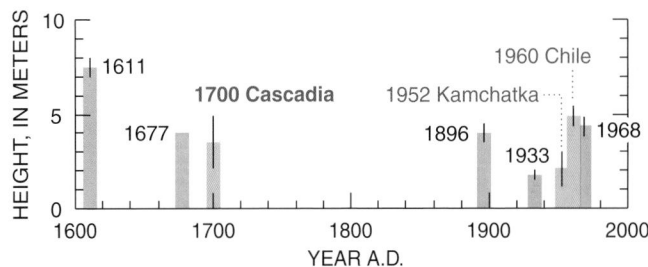

NOTABLE TSUNAMIS IN TSUGARUISHI SINCE 1600

Height along shore at south end of Miyako Bay
- Tsunami generated near Honshu
- Tsunami from distant source
- Range of estimates or measurements

1960 CHILE TSUNAMI

Miyako Bay   Flooded fields   Pines along shore near Akamae

Tsunami

Roofs of Norinowaki, weighted against typhoons

THE GRAPHED HEIGHTS of the 1952 and later tsunamis were measured soon after each event (The Central Meteorological Observatory, 1953, p. 22, 46; The Committee for Field Investigation of the Chilean Tsunami of 1960, 1961, p. 178-179; Kitamura and others, 1961a, p. 239; Kajiura and others, 1968, p. 1370, 1374). The graphed heights of earlier tsunamis are inferences from descriptions of flooding and damage (Hatori, 1995, p. 64; Tsuji and Ueda, 1995, p. 97; our pages 56-57). The 1611 tsunami caused about 150 deaths near the south end of Miyako Bay (Hatori, 1995, p. 60; Tsuji and Ueda, 1995, p. 96). Sixteen died in 1896 (Yamashita, 1997, p. 113), three in 1933 (Usami, 1996, p. 189). Katō and others (1961) describe amplification of the 1960 tsunami in Miyako Bay.

THE 1960 PHOTO, from Japan Meteorological Agency (1961, p. 339), shows a view east-northeast from the site plotted on page 56. The Akamae pines, in which people survived the 1960 tsunami, are probably similar in location to the pines shown in the picture map from 1739 (facing page).

Account in Moriai-ke "Nikki kakitome chō" 盛合家『日記書留帳』の記述

A "DIARY MEMO NOTEBOOK" of Tsugaruishi's Moriai family recounts flooding at the south end of Miyako Bay, relates it to a nearby fire, and notes the lack of an earthquake.

The water swept away houses along the bayshore and went as far inland as Kubota Crossing (columns 1-2). It caused panic in Tsugaruishi by reaching Inarinoshita, an area below Inari shrine (3). The related fire, in Kuwagasaki, destroyed about 21 houses according to hearsay (3-5). The account's author probably suspected a tsunami, for he noted that no earthquake accompanied the event (4-5).

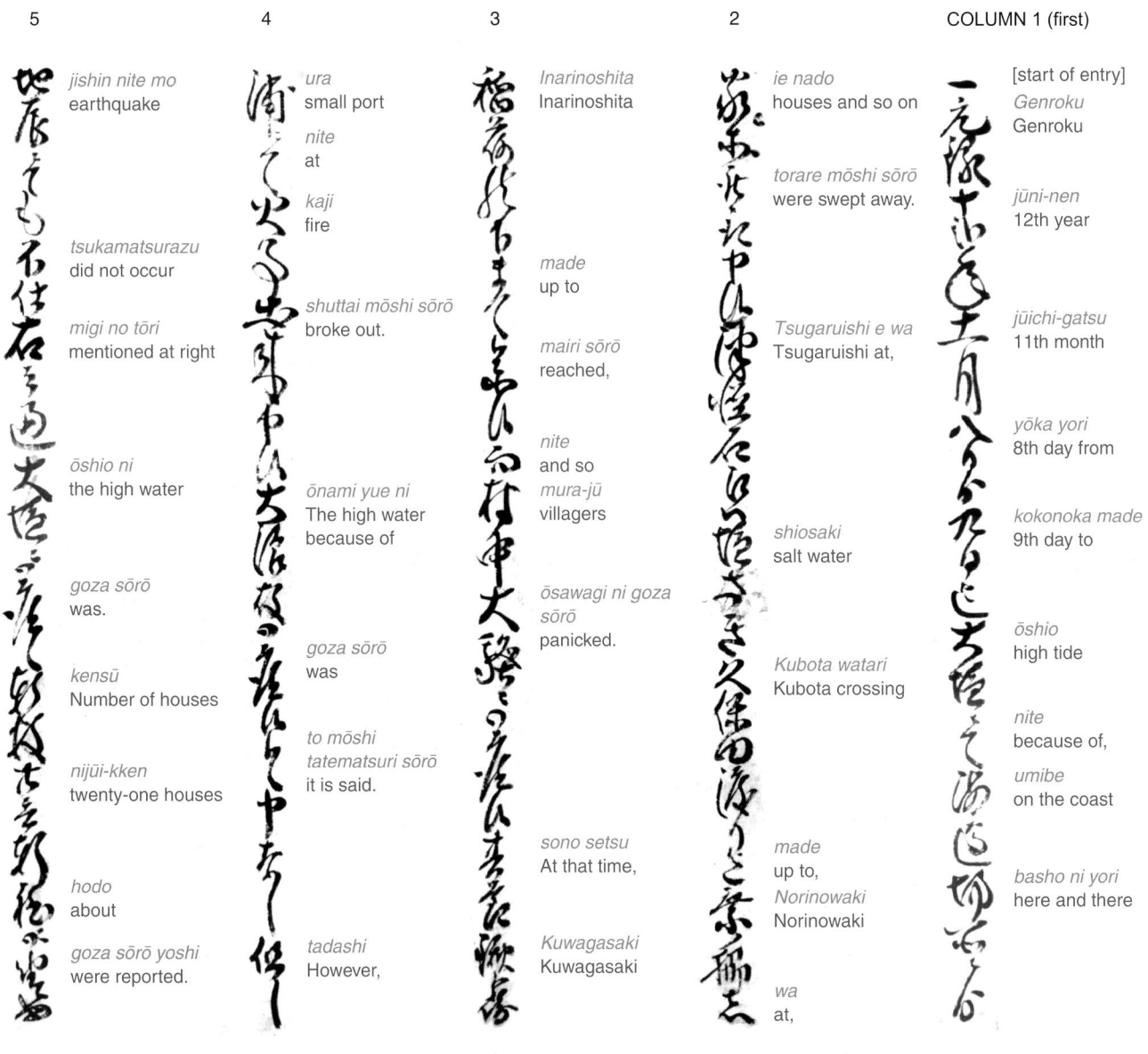

5	4	3	2	COLUMN 1 (first)
jishin nite mo earthquake	*ura* small port	*Inarinoshita* Inarinoshita	*ie nado* houses and so on	[start of entry] *Genroku* Genroku
tsukamatsurazu did not occur	*nite* at	*made* up to	*torare mōshi sōrō* were swept away.	*jūni-nen* 12th year
migi no tōri mentioned at right	*kaji* fire	*mairi sōrō* reached,	*Tsugaruishi e wa* Tsugaruishi at,	*jūichi-gatsu* 11th month
	shuttai mōshi sōrō broke out.	*nite* and so		*yōka yori* 8th day from
ōshio ni the high water		*mura-jū* villagers	*shiosaki* salt water	*kokonoka made* 9th day to
goza sōrō was.	*ōnami yue ni* The high water because of	*ōsawagi ni goza sōrō* panicked.	*Kubota watari* Kubota crossing	*ōshio* high tide
kensū Number of houses	*goza sōrō* was			*nite* because of,
nijūi-kken twenty-one houses	*to mōshi tatematsuri sōrō* it is said.		*made* up to, *Norinowaki* Norinowaki	*umibe* on the coast
hodo about		*sono setsu* At that time,		*basho ni yori* here and there
goza sōrō yoshi were reported.	*tadashi* However,	*Kuwagasaki* Kuwagasaki	*wa* at,	

5, *migi no tōri* (as at right)—Refers to material stated previously, in a column to the right (as in columns 9, 11, and 12 on p. 38).

5, *nijūi-kken hodo* (approximately 21 houses)—The houses that burned in Kuwagasaki (p. 39, column 3). Reported as hearsay.

Formal language—*mōshi sōrō* (2, 4), *mairi sōrō* (3), *goza sōrō* (3-5), *tsukamatsurazu* (5).

Sound change at word juncture—*mura-jū* for *mura-chū* (3), *nijūi-kken* for *nijūichi-ken* (5).

1, *basho ni yori*—Not fully translated. Literally, "depending on the place."

2, *ie nado*—The *nado* ("and so on") makes the *ie* plural: "houses."

2, *Norinowaki* 法之脇 —Village (maps, p. 51, 56). Probably includes the area of the houses in the foreground of the photo on p. 51. In transcribing Moriai-ke "Nikki kakitome chō," Mori (1983, p. 161) read *nori* 法 as *nori* 乗, "to ride," and he inserted a comma after this 乗. Thus in Mori's transcription, salt water "rode" to Kubota Crossing.

← NOTES. Column 1, *jūichi-gatsu* (11th month)—A mistake in copying *jūni-gatsu* (12th month). The writer repeated this mistake for a heavy snow that fell ten days after the orphan tsunami (facing page). This snowfall is securely dated in Morioka-han "Zassho" and in Hachinohe-han "Han nikki." In the latter, the heavy snow was noted independently by the *metsukesho* (inspection bureau) and by its *kanjōsho* (finance office) (Hachinohe Komonjo Benkyō-kai, 1994, p. 203).

Human error　写しまちがい

A Tsugaruishi writer miswrote the orphan tsunami's month.

ONE LUNAR MONTH separates two reported dates for a fire that destroyed some 20 houses in Kuwagasaki during a sea flood late in the year Genroku 12. Morioka-han "Zassho" dates this unusual event to the 8th day of the 12th month (p. 39, column 1); "Nikki kakitome chō," to the 8th and 9th days of the 11th month (p. 52, column 1; excerpt, right).

Errors in compiling "Nikki" probably explain this discrepancy and an adjoining one. The next entry in "Nikki kakitome chō" (p. 50), nominally for the 18th day of the 11th month, tells of heavy snow (right). Morioka-han "Zassho,"

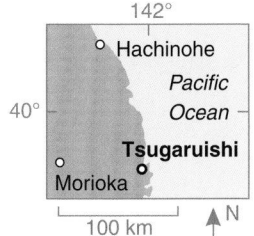

however, reports fair skies on that day and snow exactly one month later. Similarly in Hachinohe, snow fell not on 11/18 but heavily on 12/18 (first footnote, opposite). Miswriting the month of this storm, the "Nikki" compiler similarly misdated the orphan tsunami.

DATED EXACTLY ONE MONTH EARLY

SNOWSTORM

dō nen
same year

jūichi-gatsu
11th month

jūhachi-nichi
18th day

SEA FLOOD AND FIRE

Genroku
Genroku

jūni-nen
12th year

jūichi-gatsu
11th month

yōka
8th day

Social status　士農工商

A merchant family that chronicled the orphan tsunami later attained samurai rank.

THE FOUNDING WARRIORS of Edo-period Japan decreed a hereditary social order that ranked samurai above farmers, farmers above artisans, and nearly everyone above merchants. However, the samurai-led governments commonly ran up debts (p. 61), which some daimyo domains partly covered by selling samurai status to merchants. Thus in 1774 Morioka-han issued, to a prosperous merchant family from Tsugaruishi, a license that elevated them to samurai with the surname Moriai.

The family's commercial ascent began in the 1680s with loans secured by land and fishing rights. Holdings grew as borrowers failed to repay. By 1776 the Moriai held timber and shipping interests in Kuwagasaki and sake breweries in Miyako and Ōtsuchi.

The family's first samurai, Moriai Chūzaemon, reviewed records from this era of financial growth. He assembled in 1777 most of the transactions graphed at right. In that same year he annotated "Nikki kakitome chō," a "diary memo notebook" that probably originated with his grandfather, Mitsutatsu, who headed the family between 1690 and 1730.

Moriai Mitsunori headed the main branch of the Moriai family in 2004. His ancestors tracked their gains in lands and fishing rights, below.

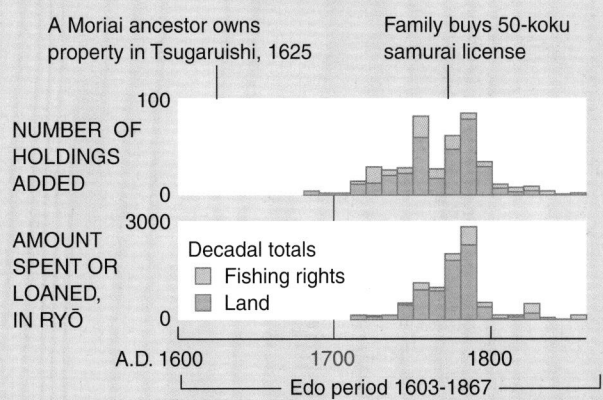

A Moriai ancestor owns property in Tsugaruishi, 1625

Family buys 50-koku samurai license

NUMBER OF HOLDINGS ADDED

AMOUNT SPENT OR LOANED, IN RYŌ

Decadal totals
Fishing rights
Land

A.D. 1600　1700　1800

Edo period 1603-1867

A MORIAI ANCESTOR, Wakasa, held property in Tsugaruishi in 1625. His descendents purchased for 410 *ryō*, in 1774, samurai status that included an annual stipend of 50 koku and official use of the family name Moriai. Iwamoto (1970, p. 98-130) describes this purchase and tabulates the family's financial records; in a later book (1979) he relates additional Moriai history. One ryō (cash) would buy about 1 koku (180 liters) of dry hulled rice (p. 71).

EIGHTEENTH-CENTURY ENTREPRENEURS paid cash to Morioka-han for "samurai status and often the rights to conduct their commerce on a restrictive basis." In 1783 "the domain issued a 'price list' for the various ranks of samurai status and privileges... [T]he right to wear swords and use a surname was 50 ryō; a promotion...to a bona-fide samurai was considerably more expensive at 620 ryō" (Hanley and Yamamura, 1977, p. 140).

Foreign waves 外国からの津波

Occasionally, a tsunami that damages Japan comes from afar.

NO EARTHQUAKE WARNED of the 1700 tsunami in Japan. No account mentions associated shaking, and two accounts note the lack of seismic warning (right).

Such orphan waves intrigued Ninomiya Saburo of the weather station in Miyako (p. 46). Soon after the 1960 Chile tsunami he matched three Edo-period tsunamis with South American earthquakes—from 1687, 1730, and 1751.

Ninomiya found no parent for the 1700 tsunami. It would remain an orphan until the 1990s (p. 93-94).

TSUGARUISHI

jishin
nite mo
earthquake

tsukamatsurazu
did not occur

MIHO

jishin
earthquake

mo
any

goza naku sōrō
did not happen

Foreign tsunamis in Japan

Earliest domestic tsunami in Japanese written history, November 26, 684 (Julian calendar)

Wave trains of unknown, probably distant source, in both cases recorded near Nakaminato

799   1420

500   1000   1500   2000

Tsunamis linked to distant earthquakes

1700 1751   1837 1877 1922 1960
1586 1687 1730   1868   1952 1964

1500   1600   1700   1800   1900   2000
YEAR A.D.

Edo period (1603-1867)

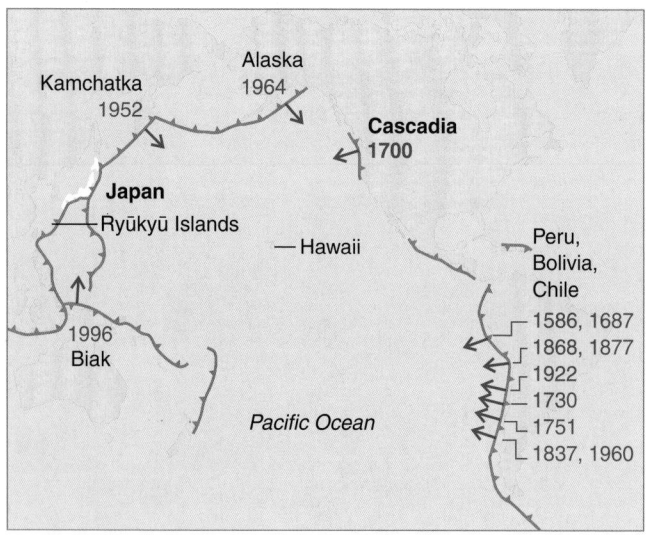

1952 **Tsunami recorded in Japan** Arrow points toward distant shores that face broad side of the tsunami's source area. Where far from its source, a tsunami tends to be largest on such shores (simulation, p. 74-75 and 99).

Subduction zone Low-angle fault between tectonic plates (p. 8, 77). Line shows upper edge; teeth point down dip. Tsunamis on map originated along subduction zones.

HONSHU

Ōtsuchi

Shinjō

1-3 m

Ōfunato

0   500 km

1586 Peru Known in Japan from southern Sanriku coast only.

1687 Peru At least 12 waves as much as 0.5 m high in Shiogama. Also known from Ryūkyū Islands (location map, lower left).

1700 Cascadia Reported from sites along 900 km of Honshu coast. Included at least seven waves and spanned parts of two days. Maximum heights probably 2-5 m (p. 48).

1730 Chile Flooded fields in Rikuzen and on Oshika Peninsula

1751 Chile In Ōtsuchi, contained seven waves and flooded house floors. In Shinjō, flooded rice fields during part of a morning.

1837 Chile Damaged rice paddies and salt works along coast between Ōfunato and Sendai (map, p. 81). Unreported from other parts of Japan but crested as high as 6 m in Hilo, Hawaii.

1952 Kamchatka Spawned by third-largest earthquake of the 20th century (M 8.8-9.0; size graphed, p. 98). Crested 1-3 m high in northern Honshu. Did not exceed 1.0 m on tide gauges (map, p. 95); however, the Miyako gauge apparently damped the 1960 Chile tsunami (footnote, p. 46).

1964 Alaska From the second-largest earthquake of 20th century (M 9.2). Maximum tide-gauged height in Japan 75 cm, at Ōfunato (p. 95). Tsunami small in Japan relative to earthquake size because its waves went mainly southeastward (arrow, map at far left).

QUOTES at top from pages 52 and 78.

ON FAR-TRAVELED TSUNAMIS see The Central Meteorological Observatory (1953, p. 39, 45-58), Ninomiya (1960), Takahashi and Hatori (1961, p. 23), Hatori (1965), Pararas-Carayannis and Calebaugh (1977), Lockridge (1985), Ōfunato Shiritsu Hakubutsukan (1990), Usami (1996), and Watanabe (1998). Perusing "Mandaiki" (p. 84) in 2002, Satake noticed a description of flooding in Shinjō, on Hōreki 1.5.2, that matches the expected arrival time of the 1751 tsunami. On tsunami directivity see Ben-Menahem and Rosenman (1972) and, for Cascadia, our pages 74-75 and 99.

JAPAN'S 684 TSUNAMI, according to the ancient chronicle "Nihongi" (or "Nihonshoki"), was "an overflowing rush of sea-water" that sank "many of the ships used for conveying tribute" (Aston, 1972, p. 366).

Ōfunato, 1960
The Chilean tsunami drove ashore the "Dai jūsan kaiun maru" ("Luck bringer no. 13").

Asahi Shimbun

Few of Japan's foreign tsunamis rival the 1700 event. In its documented Asian extent, it exceeds all other foreign tsunamis before 1868 with the exception of the South American waves of 1687 and 1751.

Japan's most ruinous foreign tsunami originated with the largest earthquake ever measured—the 1960 Chile shock of magnitude 9.5 (p. 10-11). The waves took nearly 24 hours

to reach Japan. The largest waves arrived a few hours after high tide in northern Honshu and at high tide to the south (p. 46, 83). They widely reached heights of 2-4 m and, where amplified in bays, locally crested at 5-6 m (map below). The waves caused 52 fatalities in Ōfunato (above and p. 81) and 71 deaths elsewhere in northeast Honshu. None of these losses occurred in areas of documented flooding in 1700.

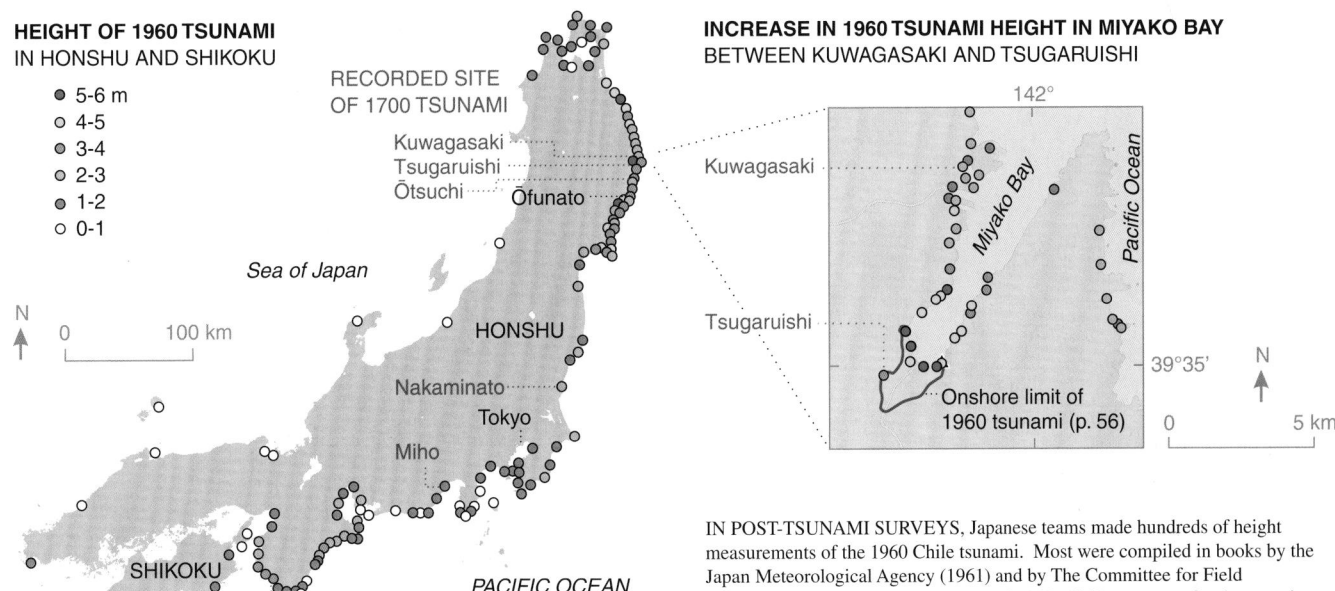

HEIGHT OF 1960 TSUNAMI
IN HONSHU AND SHIKOKU

- 5-6 m
- 4-5
- 3-4
- 2-3
- 1-2
- 0-1

RECORDED SITE OF 1700 TSUNAMI

Kuwagasaki
Tsugaruishi
Ōtsuchi
Ōfunato

Sea of Japan

N

0   100 km

HONSHU

Nakaminato
Tokyo
Miho

SHIKOKU

PACIFIC OCEAN

Tanabe and Shinjō

INCREASE IN 1960 TSUNAMI HEIGHT IN MIYAKO BAY
BETWEEN KUWAGASAKI AND TSUGARUISHI

142°

Kuwagasaki

Miyako Bay

Pacific Ocean

Tsugaruishi

39°35'

N

Onshore limit of 1960 tsunami (p. 56)

0   5 km

IN POST-TSUNAMI SURVEYS, Japanese teams made hundreds of height measurements of the 1960 Chile tsunami. Most were compiled in books by the Japan Meteorological Agency (1961) and by The Committee for Field Investigation of the Chilean Tsunami of 1960 (1961)—sources for the overview map at left and most details above. The height estimate for Tsugaruishi is based on tsunami limits identified by eyewitnesses interviewed in 1999 (p. 56-57).

Tsunami size 津波の高さ

The 1700 tsunami probably grew to heights of five meters in Miyako Bay.

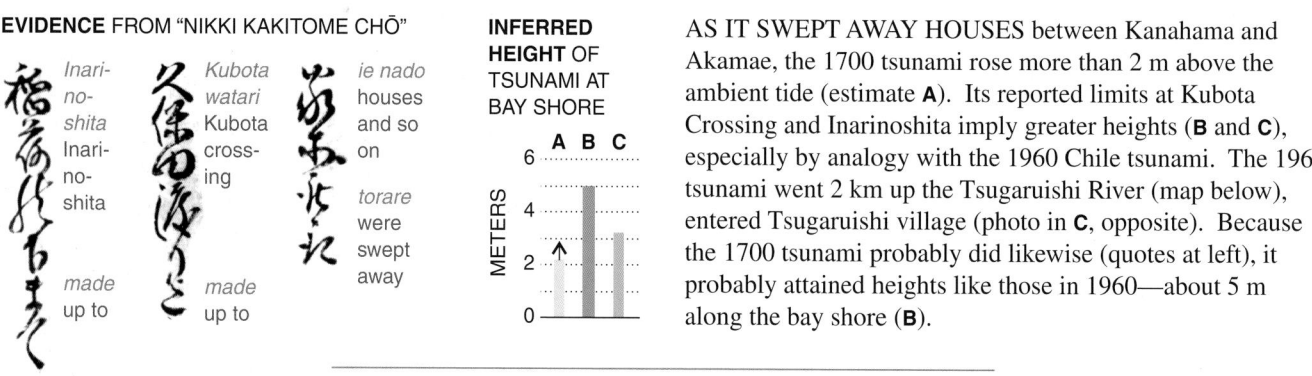

EVIDENCE FROM "NIKKI KAKITOME CHŌ"

Inari-no-shita
Inari-no-shita

made up to

Kubota watari Kubota crossing

made up to

ie nado houses and so on

torare were swept away

INFERRED HEIGHT OF TSUNAMI AT BAY SHORE

```
METERS
6
4
2
0
   A B C
```

AS IT SWEPT AWAY HOUSES between Kanahama and Akamae, the 1700 tsunami rose more than 2 m above the ambient tide (estimate **A**). Its reported limits at Kubota Crossing and Inarinoshita imply greater heights (**B** and **C**), especially by analogy with the 1960 Chile tsunami. The 1960 tsunami went 2 km up the Tsugaruishi River (map below), entered Tsugaruishi village (photo in **C**, opposite). Because the 1700 tsunami probably did likewise (quotes at left), it probably attained heights like those in 1960—about 5 m along the bay shore (**B**).

SETTING OF THE 1700 TSUNAMI, AND OBSERVED LIMITS AND HEIGHTS OF THE 1960 TSUNAMI

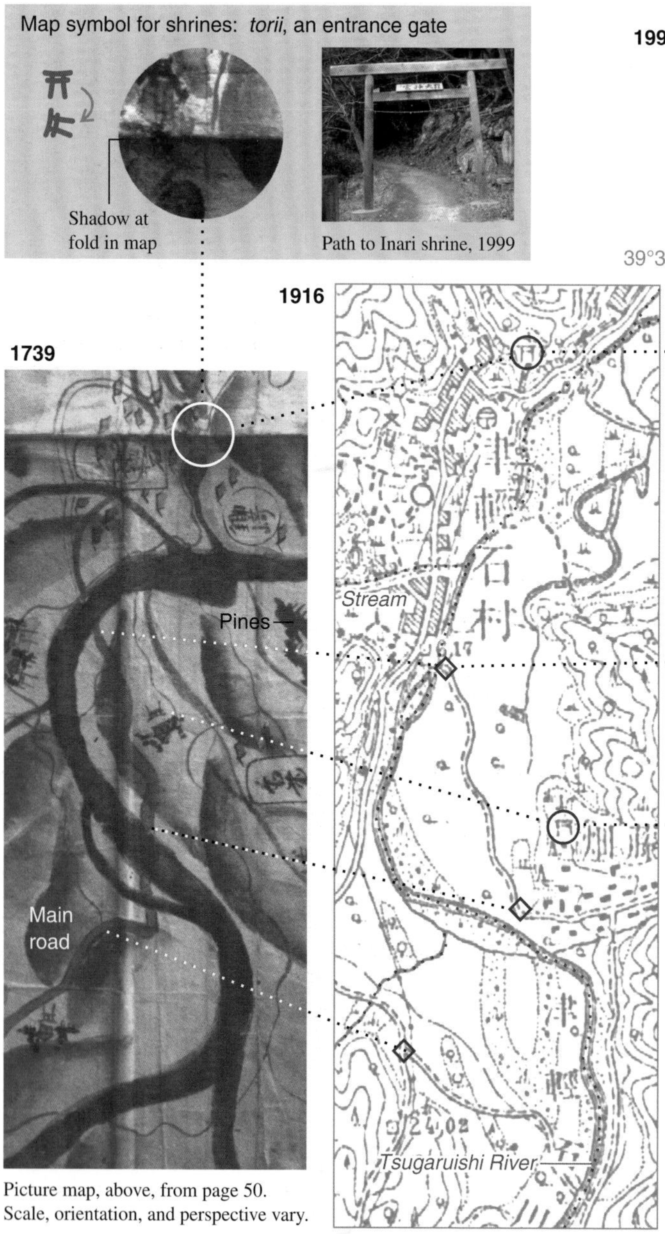

Map symbol for shrines: *torii*, an entrance gate

Shadow at fold in map

Path to Inari shrine, 1999

1739

Main road

Pines

Stream

1916

Tsugaruishi River

Picture map, above, from page 50. Scale, orientation, and perspective vary.

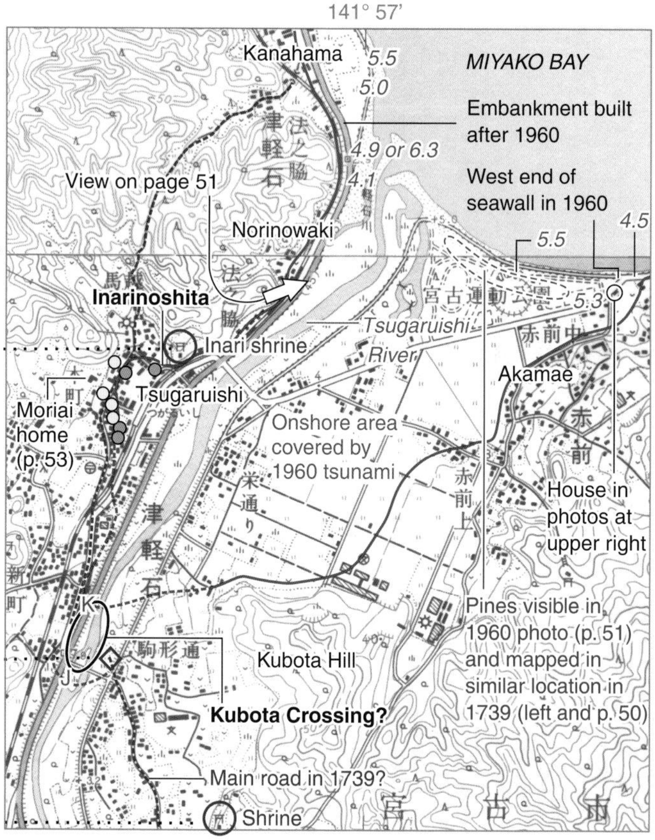

141° 57'

1996

MIYAKO BAY

Kanahama 5.5
           5.0

Embankment built after 1960

View on page 51 4.9 or 6.3
                 4.1

West end of seawall in 1960

Norinowaki

4.5

39°35'

馬城
法之脇

Inarinoshita

5.5

Inari shrine

Tsugaruishi River

宮古運動   5.3

Tsugaruishi

Moriai home (p. 53)

Akamae

Onshore area covered by 1960 tsunami

House in photos at upper right

Kubota Hill

Pines visible in 1960 photo (p. 51) and mapped in similar location in 1739 (left and p. 50)

駒形通

Kubota Crossing?

Main road in 1739?

Shrine

5.3 Height of 1960 tsunami (m above TP)
⌐ Landward limit of 1960 tsunami—Dashed where sources differ (J and K, below). Control points from interviews in 1999:
○ above limit, ● below limit

N

Scale 1:25,000
0   1 km

Contour intervals 10 m (1996) and 20 m (1916)

1960 TSUNAMI heights from The Committee for the Field Investigation of the Chilean Tsunami of 1960 (1961, p. 178-179) and Japan Meteorological Agency (1961, p. 119). TP, Tokyo Peil, a datum near mean sea level (p. 46). Landward limits: K, Kon'no (1961, p. 22) and Kitamura and others (1961a, p. 239); J, Japan Meteorological Agency (1961, p. 119).

BASE MAPS from Kokudo Chiriin (Geographical Survey Institute), Miyako and Tsugaruishi 1:25,000, 1996; and Rikuchi Sokuryōbu, Miyako 1:50,000, 1916.

A Minimum height inferred from loss of houses beside Miyako Bay

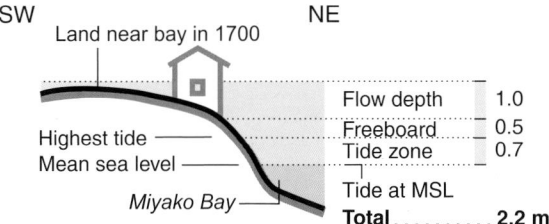

SW   NE

Land near bay in 1700

Highest tide
Mean sea level
Miyako Bay

Flow depth	1.0
Freeboard	0.5
Tide zone	0.7
Tide at MSL	
Total . . . . **2.2 m**	

ASSUMPTIONS

Flow depth Tsunami crested 1 m deep where it destroyed houses.

Freeboard To avoid flooding during storm tides—and perhaps with the 1677 tsunami in recent memory (p. 51)—villagers sited their houses no less than 0.5 m above highest astronomical tide.

Tide zone The highest astronomical tide in 1700 was 0.7 m above mean sea level, by analogy with modern tides recorded in Kuwagasaki Harbor (footnote, p. 48).

Tide stage No correction attempted because "Nikki" gives dates of flooding but not its time.

High-water line 5.3 m above TP

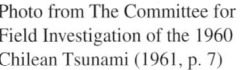

Plaque marks line

Photo from The Committee for Field Investigation of the 1960 Chilean Tsunami (1961, p. 7)

The 1960 Chile tsunami gutted this house in Akamae, 75 m inland from Miyako Bay. In 1999 a sign marked the 1960 high-water line.

B More realistic height inferred from inundation to Kubota Crossing

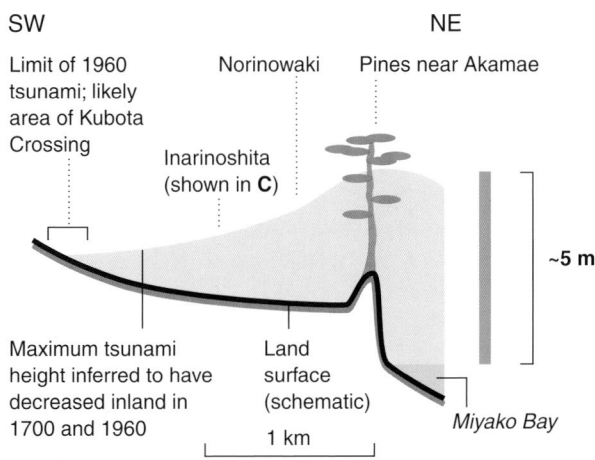

SW   NE

Limit of 1960 tsunami; likely area of Kubota Crossing

Norinowaki   Pines near Akamae

Inarinoshita (shown in **C**)

~5 m

Maximum tsunami height inferred to have decreased inland in 1700 and 1960

Land surface (schematic)

Miyako Bay

1 km

ASSUMPTIONS

Kubota Crossing Denotes a ferry where the area's main Edo-period road intersected the Tsugaruishi River. Two such crossings appear on the picture map from 1739 (far left). The one nearer Miyako Bay coincides with the 1960 tsunami's upriver limit, as judged from points shared with later maps (linked, facing page). "Kubota Hill" is the local name for high ground that overlooks this area, 2 km inland from the bay.

1960 analogy If the 1700 and 1960 tsunamis had similar inland limits, they probably reached similar heights at the south shore of Miyako Bay. On that shore the 1960 tsunami crested at 4.5-5.5 m.

Inland decline Both tsunamis probably decreased inland in maximum height. The 1960 maximum probably descended from 5.5 m at the bay to 3.5 m at Inarinoshita, where the water crested about 1.5 m deep on land 2 m above TP (likely site shown in photo below). Relative to TP, the 1960 maximum also descended landward at Ōtsuchi and Shinjō (p. 65, 89).

C Height inferred from inundation to Inarinoshita, adjusted for tectonic subsidence

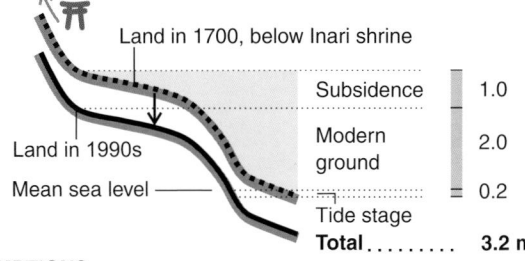

Land in 1700, below Inari shrine

Subsidence	1.0
Modern ground	2.0
	0.2
Tide stage	
Total . . . . **3.2 m**	

Land in 1990s
Mean sea level

ASSUMPTIONS

Modern ground Inarinoshita, now Inarigashita, refers to the area at right. (This area lies below, *shita*, a hill between Tsugaruishi and Norinowaki on which a shrine to Inari, a Shinto god, has stood since 1635 or earlier.) Flat ground in the photo is 2.0 m above TP.

Subsidence Relative to the sea, about 1 m since 1700 (see p. 65)

Tide stage 0.2 m below 1700 mean sea level (p. 83)

Height **C** from Tsuji and others (1998). Tsugaruishi fishermen went to "Inari Hill" in 1635 for divination of the year's catch (Iwamoto, 1970, p. 21). Moriai Miya, interviewed in 1999 (p. 107), identified a 1960 high-water line near the site site at right, for which a modern municipal map gives a height of 2 m TP.

Inari shrine

The 1700 tsunami probably flooded the site of the Tsugaruishi neighborhood now known as Inarigashita. Here, the 1960 tsunami flowed about 1.5 m deep and destroyed a house.

Ōtsuchi 大槌

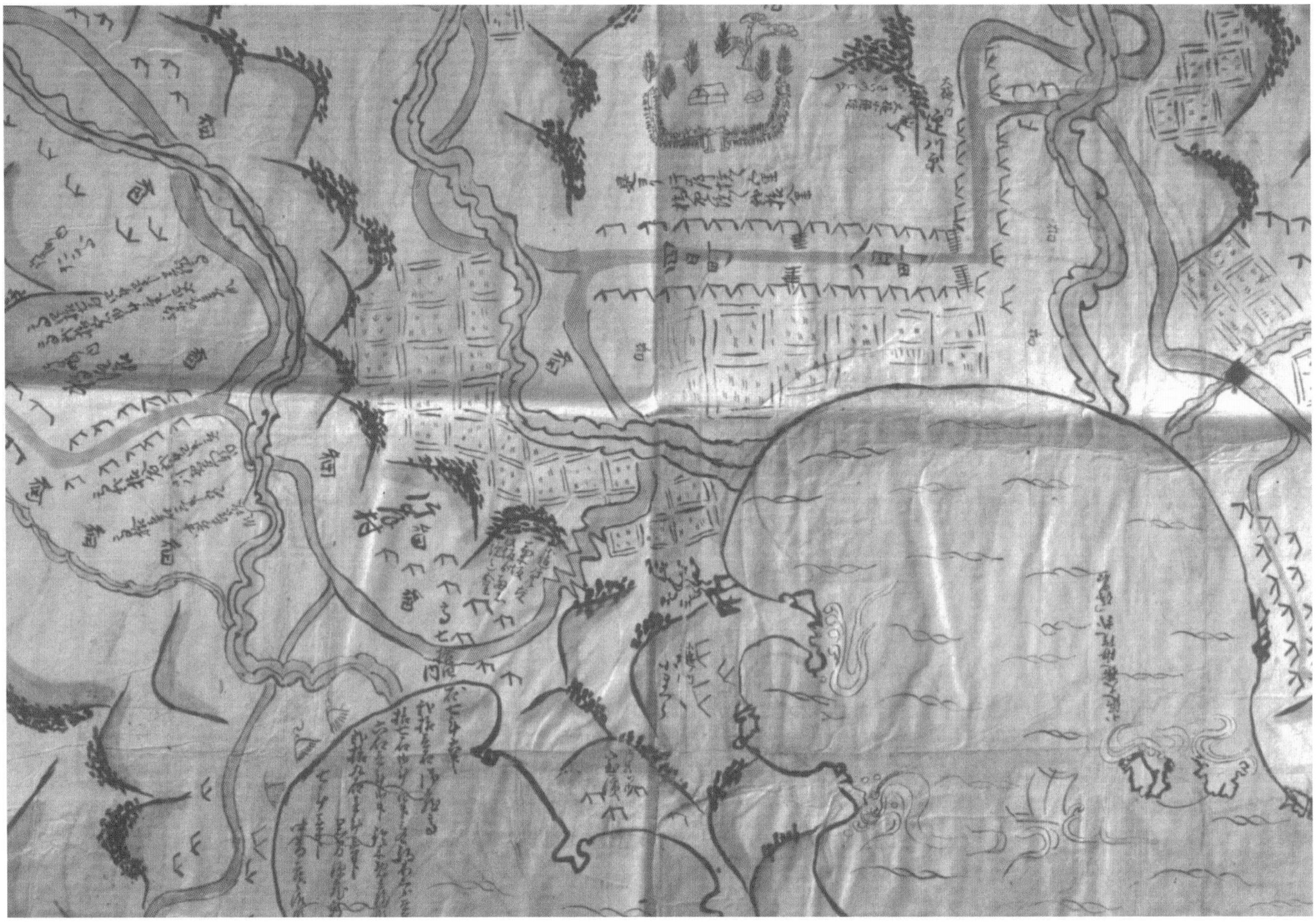

On a tax map probably made in 1730, Ōtsuchi's houses line a road between bayside paddies and the district magistrates' office. To the southwest, smoke rises from kilns.

Approximate north ↑

Ōtsuchi magistrates stationed here, in a government office building (*o-yakuya* 御役屋), sent a report on the 1700 tsunami to Morioka castle. A later Ōtsuchi magistrate prepared a summary from which a Japanese earthquake historian would learn of the 1700 tsunami by 1943 (p. 62).

The kilns may mark the area of salt evaporators reportedly damaged by the 1700 tsunami. Notes below kilns identify recipients of rice-tax revenues.

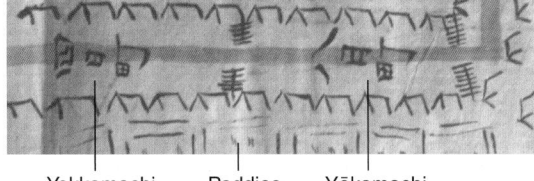

Yokkamachi   Paddies   Yōkamachi

Ōtsuchi's main street probably escaped flooding by the 1700 tsunami and also by the 1960 tsunami. Lining the street were neighborhoods named for their market days. The 1677 tsunami, of nearby source, entered 20 houses in the eighth-day neighborhood, Yōkamachi.

THE PICTURE MAP, conserved at Morioka-shi Chūō Kōminkan (p. 44), is probably tied to tax records from about 1730, according to Konishi Hiroaki, the Kōminkan documents librarian (interviewed 1999). Text beside kilns describes division of 74 *koku* (about 13,000 liters) of rice among Morioka-han (21 koku) and three samurai (17, 6, and 29 koku). In the fourth-day neighborhood (四日町 Yokkamachi), market days ended in four (4th day, 14th day, 24th day); likewise in the eighth-day neighborhood (八日町 Yōkamachi), markets were open on the 8th, 18th, and 28th.

ŌTSUCHI STATISTICS for 1803: 1465 persons, 273 houses, 42 boats (Takeuchi, 1985a, p. 172).

Main points

The sea invaded Ōtsuchi the same date and hour as it did 30 km to the north, in Kuwagasaki (p. 43, 72).

The flooding damaged paddies, two houses, and two salt-evaporation kilns (p. 60). This damage, though small, was reported to Edo, perhaps to help justify financial relief from the Tokugawa shogunate (p. 61).

An earthquake historian included this flooding in an earthquake catalog issued in 1943 (p. 62).

The flooding in 1700 probably stopped short of Ōtsuchi's main Edo-period street. The 1751 Chile tsunami reportedly crossed this street, but the 1960 Chile tsunami did not (p. 64).

Because of puzzling regional subsidence, places covered by the 1700 tsunami in Ōtsuchi may now stand a meter lower, relative to the sea, than they did in 1700 (p. 65).

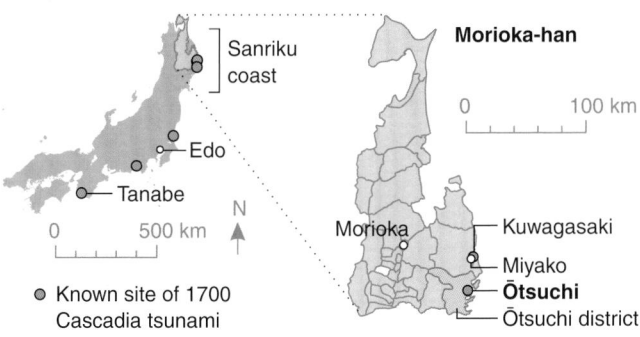

● Known site of 1700 Cascadia tsunami

Setting

Nestled between hills and bayside paddies, Edo-period Ōtsuchi stretched along a road between two river mouths. Houses flanked both sides of the street. In a side valley stood the office of a magistrate, or magistrates, who administered the Ōtsuchi district of Morioka-han.

Documents

Morioka-han "Zassho" provides the main account of the 1700 tsunami in Ōtsuchi. Like the entry about the tsunami in Kuwagasaki (p. 36), it is based on a report from coastal magistrates (p. 44). The report from Ōtsuchi probably reached Morioka castle the day after the report on Kuwagasaki arrived from Miyako (p. 60).

The 1700 tsunami killed no person or horse in Ōtsuchi, according to "Ōtsuchi kokon daidenki," a chronological record of the Ōtsuchi magistrates' office. A secondary source, "Daidenki" contains material from 1596 to 1796 that was compiled and edited in Ōtsuchi by Ogawa Magobei Yoshiyasu (1735-1820). The oldest surviving version was copied from Ogawa's compilation. The compiler or the copyist wrote the 1700 tsunami's month and hour but neglected its day. Before earthquake historians found the "Zassho" account, this omission in "Daidenki" obscured the link between the 1700 tsunami in Ōtsuchi and the flooding of similar character in Tanabe (p. 62).

Sanriku ō-tsunami
Sanriku great tsunamis

dekishisha seirei
drowned-persons' souls

kuyōtō
monument for prayer

A tsunami memorial, for victims in 1896 and 1933, stands in the cemetery of Kōganji Temple, Ōtsuchi.

Other tsunamis

Tsunamis generated off northeast Honshu devastated Ōtsuchi in 1611, 1896, and 1933. Deaths from the 1611 waves totaled about 800 in Ōtsuchi and vicinity. In the town of Ōtsuchi alone, the 1896 and 1933 tsunamis took 600 and 61 lives, respectively. An inscription on the back of a memorial stone, above, further states that the town lost more than 600 houses to each of these latter tsunamis.

Lesser near-source tsunamis reached heights of several meters in Ōtsuchi in 1677, 1793, 1856, and 1968. The 1677 tsunami covered the floor in 20 of 60 houses in the Yōkamachi neighborhood along the town's main street.

Among tsunamis of remote origin, 1751 Chile may have reached the farthest into Ōtsuchi. It entered both the Yokkamachi and Yōkamachi neighborhoods, flooding a dozen houses. The 1960 Chile tsunami approached 4 m in height along the bayshore south of town. Its crest descended onshore to the tsunami's limit near the 2 m topographic contour, seaward of the main Edo-period street (p. 64-65).

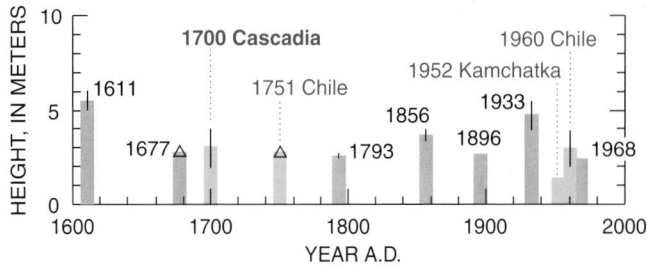

NOTABLE TSUNAMIS IN ŌTSUCHI SINCE 1600

■ Tsunami generated near Honshu
■ Tsunami from distant source
△ 1677 and 1751—Flooding went farther inland than in 1700.
| Range of estimates or measurements

TSUNAMI MEMORIAL mapped on page 65.
TSUNAMI HEIGHTS and deaths: **1611**, Hatori (1995, p. 60, 62). **1677**, Tsuji and Ueda (1995, p. 102). Height excludes correction for tectonic change in land-level (p. 65). **1700**, our pages 64-65. **1751**, height by analogy with 1677. On flooding in 1751, see Ninomiya (1960, p. 20) and Watanabe (1998, p. 217-218). **1793**, **1856**, **1896**, and **1933**, Tsuji and Ueda (1995, p. 103). Deaths in 1896, Yamashita (1997, p. 113). Deaths and 5.5 m height in 1933, Watanabe (1998, p. 115, 118). **1952**, The Central Meteorological Observatory (1953, p. 46). **1960**, our page 65. **1968**, Kajiura and others (1968, p. 1373).

Account in Morioka-han "Zassho" 盛岡藩『雑書』の記述

HIGH WATER came to the small port of Ōtsuchi at a time equivalent to midnight (column 1; clock, p. 43). Damaged were rice paddies and vegetable fields seaward of Ōtsuchi's main street (2). In addition, the Ōtsuchi magistrate's office learned of damage to two houses and two salt kilns (2-3). All this news was forwarded to Edo (4).

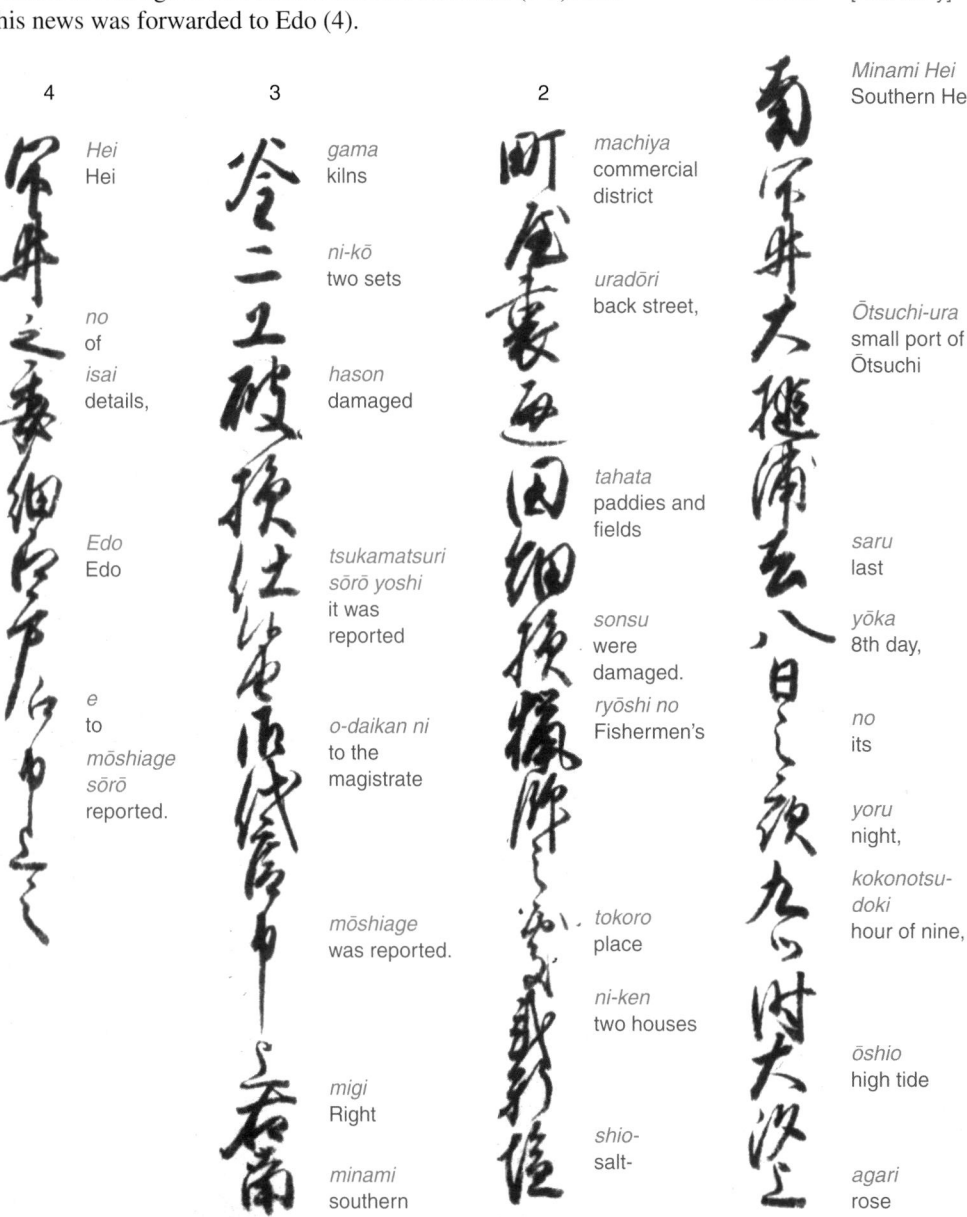

COLUMN 1

— [new entry]

Minami Hei Southern Hei

Ōtsuchi-ura small port of Ōtsuchi

saru last

yōka 8th day,

no its

yoru night,

kokonotsu-doki hour of nine,

ōshio high tide

agari rose

Column 2

machiya commercial district

uradōri back street,

hason damaged

tsukamatsuri sōrō yoshi it was reported

o-daikan ni to the magistrate

mōshiage was reported.

tokoro place

ni-ken two houses

shio- salt-

Column 3

gama kilns

ni-kō two sets

hason damaged

migi Right

minami southern

Column 4

Hei Hei

no of

isai details,

Edo Edo

e to

mōshiage sōrō reported.

HEADNOTE

jūgo-nichi 15th day

hare kaze fair, windy —

Kichibe'e, Kyūbe'e, Jinzaemon [karō, p. 44]

Daily headnotes in Morioka-han "Zassho" probably date the receipt of information at Morioka castle. Thus the Miyako magistrates' report of the 1700 tsunami in Kuwagasaki likely reached the castle on the 14th day, 12th month, 12th year of the Genroku era (p. 44)—on February 2, 1700 (p. 40, 42). The Ōtsuchi report came a day later, as dated by its headnote, above. Neither report took more than seven days to reach Morioka.

Ōtsuchi magistrates (to the *daikan*). In *hason tsukamatsuri sōrō yoshi*, the magistrates offer the news to their own superiors in Morioka (to the karō Kichibe'e, Kyūbe'e, and Jinzaemon, p. 44).

3, *migi*—Mentioned in a previous column, at right.

4, *Edo e mōshiage sōrō*—Officials in Morioka forwarded the news to the domain's officials in Edo. See facing page.

4, *e*—Pronounced and written as *e*, signifies "to."

Sound change at word juncture—*doki* for *toki* (1), *dōri* for *tōri* (2), *gama* for *kama* (3).

2, *uradōri*—Behind (*ura*), street (*tōri*); back street?

2, *ken*—Counter for houses (also used p. 39, columns 3 and 4; p. 52, column 5).

2-3, *shiogama*—Literally, salt kettle. Interpreted as "kilns" because the map from ca. 1730 shows kilns on a beach near Ōtsuchi (p. 58).

3, *hason tsukamatsuri sōrō yoshi o-daikan ni mōshiage*—A combination of formal and humble language for three social levels involved in reporting the damage to houses and kilns. In *o-daikan ni mōshiage*, the verb *mōshiage* implies that commoners reported the damage to the

← NOTES. Columns 1 and 3-4, *Minami Hei*—Mutsu province (labeled, p. 32), contained 54 counties, or *gun* (Suruga province contained seven, p. 31). Morioka-han administered Hei-gun as two districts, Miyako-dōri in the north and Ōtsuchi-dōri in the south (map, p. 44).

1, *ura*—Small port; unlike Miyako, lacks shipping route on shogunal map from 1702 (red line, p. 33).

1, *ōshio*—Parsed on page 40.

2, *machiya*—Probably refers to the neighborhoods *yōka-machi* and *yokka-machi* (p. 58).

Report to Edo 江戸への報告

Why forward, to the shogun's capital, the particulars of minor damage to a minor port?

SEVERAL REPORTS of the 1700 tsunami make their immediate purpose clear. Magistrates justify an allocation of rice and a request for wood in Kuwagasaki. Other magistrates certify the sinking of 28 tons of rice off Nakaminato. A headman wonders about stealth waves in Miho. A mayor in Tanabe expresses concern about the flooding of a nearby storehouse that belongs to a branch of Japan's ruling Tokugawa family.

Left unstated, in column 4 at left, is why samurai in Morioka castle forwarded to Edo—the shogun's bustling capital—details on small losses from a natural disturbance to a remote shore. We speculate that officials of Morioka-han kept track of natural disasters in hopes of financial relief from the Tokugawa shogunate.

Morioka-han spent heavily to comply with Tokugawa edicts. Through most of the Edo period, the shogunate required daimyo—some 260 land barons, including the distant Nambu governor of Morioka-han—to reside alternate years in Edo. This required residence consumed over half the tax income of Morioka-han. The Nambu governor would journey between Morioka and Edo, 546 km by road, with a showy entourage of some 250 persons and 100 horses. While in Edo, he would reside in a mansion near the shogun's castle (map, lower right).

The shogunate allowed Morioka-han to cancel this journey after poor harvests in 1695. Reportedly starving that year were 34,000 people—ten percent of the domain's population. The domain's next famine, in progress at the time of the 1700 tsunami, resulted from frost and rain during the 1699 growing season. The domain's records state that 27,186 people suffered from hunger that year. Such circumstances may have spurred domain officials to document natural disasters as minor as the 1700 tsunami in Ōtsuchi.

THE NUMBER OF DAIMYO DOMAINS, or han, stood at 241 in 1688 and reached 262 by 1720 (Bolitho, 1991, p. 201). Each han had an expected annual production worth at least 10,000 koku of rice (on koku and hyō, see page 71).

IN MORIOKA-HAN, according to figures for internal use (naidaka; Tsukahira, 1966, p. 82), the expected yield was 241,922 koku (Mori, 1963, p. 921) on the basis of surveys in 1666-1683 (Kambun 6 to Ten'na 3). For the required residence in Edo (sankin kōtai), the governor's entourage journeyed 139 ri between Morioka and Edo (Hosoi, 1988, p. 79). The domain's yearly costs of sankin kōtai in the 17th century have been estimated conservatively as 46,500 ryō (Hanley and Yamamura, 1977, p. 130-131). To help cover these expenses, Morioka-han levied special taxes on its samurai (1.5 ryō per 100 koku of stipend) and on its commoners (total yearly revenues about 8,300 ryō).

THE 34,000 REPORTEDLY STARVING after the 1695 harvest compares with 334,887, the commoner population recorded by Morioka-han for 1695 (Mori, 1963, p. 642). Mori (1972, p. 125) associates the poor harvest in 1695 with this famine and with the cancellation of sankin kōtai. He lists a shortfall of 100,000 hyō in 1695 and 77,320 hyō in 1699. At 0.4 koku per hyō, the shortfall in 1699 amounts to nearly 31,000 koku, or about one-eighth the domain's naidaka. The Tokugawa shogunate (the bakufu) commonly helped daimyo domains after disasters: "By far the most common type of assistance took the form of loans [that were] reserved for the emergencies of which Tokugawa Japan seemed to produce an inordinate number. [W]henever a crop was ruined, a castle damaged, or an Edo mansion destroyed, the bakufu could be relied on for aid" (Bolitho, 1991, p. 202).

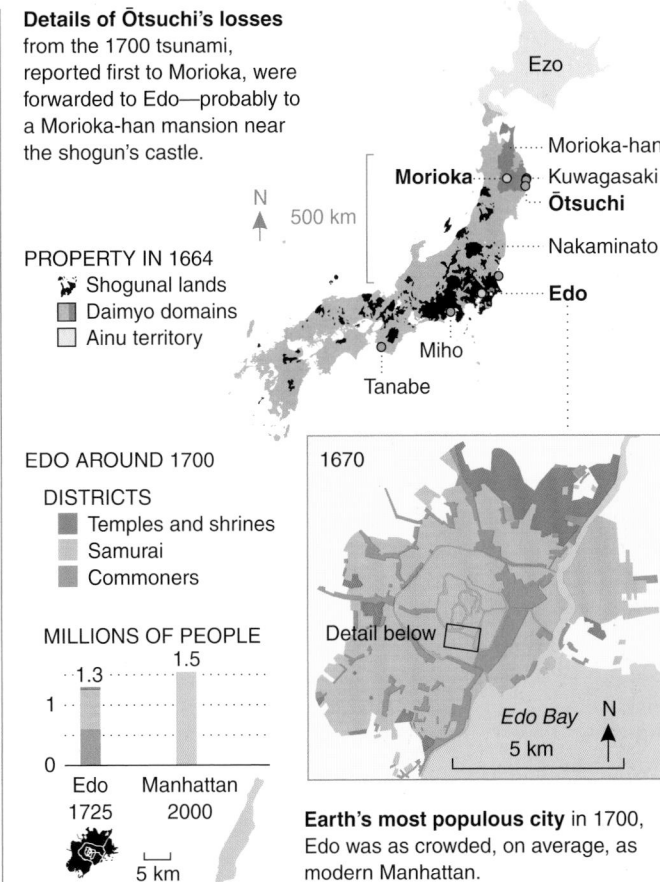

Details of Ōtsuchi's losses from the 1700 tsunami, reported first to Morioka, were forwarded to Edo—probably to a Morioka-han mansion near the shogun's castle.

PROPERTY IN 1664
- Shogunal lands
- Daimyo domains
- Ainu territory

EDO AROUND 1700

DISTRICTS
- Temples and shrines
- Samurai
- Commoners

MILLIONS OF PEOPLE

	1.3	1.5

Edo 1725 — Manhattan 2000

Earth's most populous city in 1700, Edo was as crowded, on average, as modern Manhattan.

1670 — Detail below — Edo Bay N — 5 km

CENTRAL EDO IN 1684

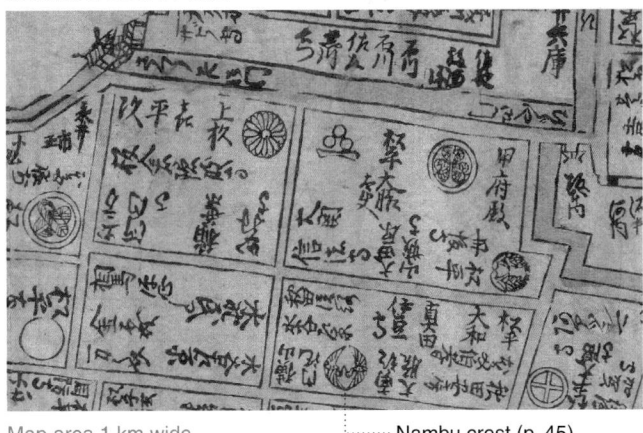

Map area 1 km wide ········ Nambu crest (p. 45)

Daimyo mansions, here marked by family crest, adjoined moats of central Edo. A corner of the shogun's castle grounds is at upper left.

LANDHOLDINGS IN 1664 from Totman (1967, map 4). The Ainu, native people, held most of Ezo (now Hokkaido) until 1800 (Walker, 2001).

ON EDO population, mansions, and maps see Naitō and Hozumi (1982; 2003, p. 104, 108, 117, 178). The main mansion (kami-yashiki) of Morioka-han stood on a five-acre estate—a parcel of 6013 tsubo (Mori, 1963, p. 360), or two hectares. The lower map is a detail from "Eiri Edo ōezu" (p. 106), courtesy of East Asian Library, University of California, Berkeley.

Collected writings 史料集

Modern recognition of the 1700 tsunami in Japan began with a teacher's mimeographed anthology of historical earthquakes.

MUSHA KINKICHI (1891-1962), an educator and geographer, collected accounts of historical earthquakes on behalf of the Earthquake Research Institute in Tokyo. He began this work in 1928 and continued it into the 1940s.

Because his mandate included events possibly related to earthquakes, Musha noted reports of high water at Ōtsuchi and Tanabe from the year 1700. Musha summarized them side-by-side in a collection brought out first as a wartime mimeograph and printed later as one of the green hardbound volumes at right.

Most other accounts of the 1700 tsunami went undiscovered until the 1970s and 1980s, when a new generation of earthquake historians began mining Japan's old documents. A project led by Usami Tatsuo and Ueda Kazue produced a 16,812-page anthology, "Shinshū Nihon jishin shiryō" ("Newly collected materials on historical earthquakes in Japan;" photo, p. 123). Its first 21 volumes appeared between 1981 and 1994.

The Musha and "Shinshū shiryō" anthologies together identify 45,698 Japanese earthquakes from the years A.D. 416 to 1872. Most date from the Edo period (1603-1867), when record-keeping first flourished outside the nation's capitals (graphs, facing page). Coincidence with the Edo period thus helped the 1700 tsunami enter written history.

MUSHA KINKICHI AND HIS ANTHOLOGY

Contents first published
1949   1943-1941

MUSHA'S SUMMARY OF ACCOUNTS OF THE 1700 TSUNAMI

Ōtsuchi | Tanabe

元禄十二年（西暦一七〇〇）十二月陸中大槌津浪アリ。
※（大槌記録抄）
この年極月夜九ツに大汐さし、海辺大分驚さす。
馬怪我なし
人

元禄十二年十二月八日（西暦一七〇〇二七）
※田辺町大帳（一）伊○紀
十二月八日潮水非常に増長
ナレリ。
紀伊國潮汐常三異

1700

8th day
— Known to Musha from Tanabe only

Y—Year, Genroku 12
M—12th month

MODERN DISCOVERY OF THE ORPHAN TSUNAMI OF 1700

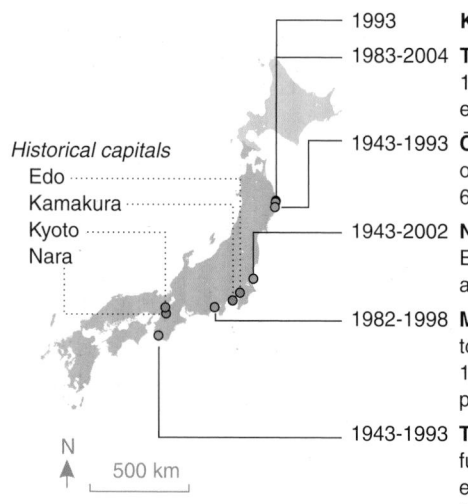

Historical capitals
Edo
Kamakura
Kyoto
Nara

N
500 km

1993 **Kuwagasaki** "Shinshū Nihon jishin shiryō" quotes Morioka-han "Zassho" (p. 38-39).

1983-2004 **Tsugaruishi** A regional historican transcribes the Moriai-ke "Nikki kakitome chō" account in 1983. The transcription is quoted ten years later in "Shinshū Nihon jishin shiryō." In 2004, earthquake historians view the 18th-century source and confirm a copyist's error (p. 50-53).

1943-1993 **Ōtsuchi** Musha Kinkichi, in his anthology from 1943, cites "Ōtsuchi kokon daidenki," which omits the day of the flooding (above). This missing detail survives in Morioka-han "Zassho" (p. 60), as quoted in "Shinshū Nihon jishin shiryō" in 1993.

1943-2002 **Nakaminato** A prefectural fisheries association prints the Nakaminato account in 1943. Earthquake historians cite this secondary source In 1996. In 2002, Satake traces the account to a family's collection of documents on Edo-period shipwrecks (p. 66).

1982-1998 **Miho** "Miho-mura yōji oboe" contains an entry on the earthquake and tsunami of 1703 alludes to the 1700 tsunami as "Rabbit Year waves" (p. 76-77). This flashback, collected by Tsuji in 1982, appears in a 1989 volume of "Shinshū Nihon jishin shiryō." Ueda, in 1998, notices that the preceding entry in "Miho-mura yōji oboe" describes the 1700 waves in detail (p. 78-79).

1943-1993 **Tanabe** Musha's 1943 anthology summarizes the account in "Tanabe-machi daichō" (above; full text on our page 86). In 1981, a parallel account from "Mandaiki" (p. 84) appears in an earthquake anthology. The 1993 volume of "Shinshū Nihon jishin shiryō" also contains the "Mandaiki" version.

DURING THE FIREBOMBING of Tokyo in 1945, the manuscript for Musha's 1949 volume awaited war's end in a galvanized box 3 m underground. Musha worked for the military geology branch of the U.S. occupation forces from 1949 to 1960. His collection of earthquake accounts built on previous work, much of it by Tayama Minoru, who issued a two-volume anthology in 1904 (Usami, 1979a, b).

USAMI (1996) summarizes descriptions of more than 300 earthquakes that struck Japan between 416 and 1872. A parallel summary for tsunamis was compiled by Watanabe (1998). Tsuji and others (1998, p. 2-4) trace the origins of accounts of the 1700 tsunami.

A PREFECTURAL FISHERIES ASSOCIATION published the Nakaminato account in a volume edited by Ōuchi (1943).

THE COLUMNS OF JAPANESE TEXT above are excerpted from Mombushō Shinsai Yobō Hyōgikai (1943, p. 25; see also our page 112). The photo of Musha Kinkichi, undated, was provided by his family through Matsu'ura Ritsuko.

JAPANESE EARTHQUAKES A.D. 600-1872, IDENTIFIED IN OLD DOCUMENTS

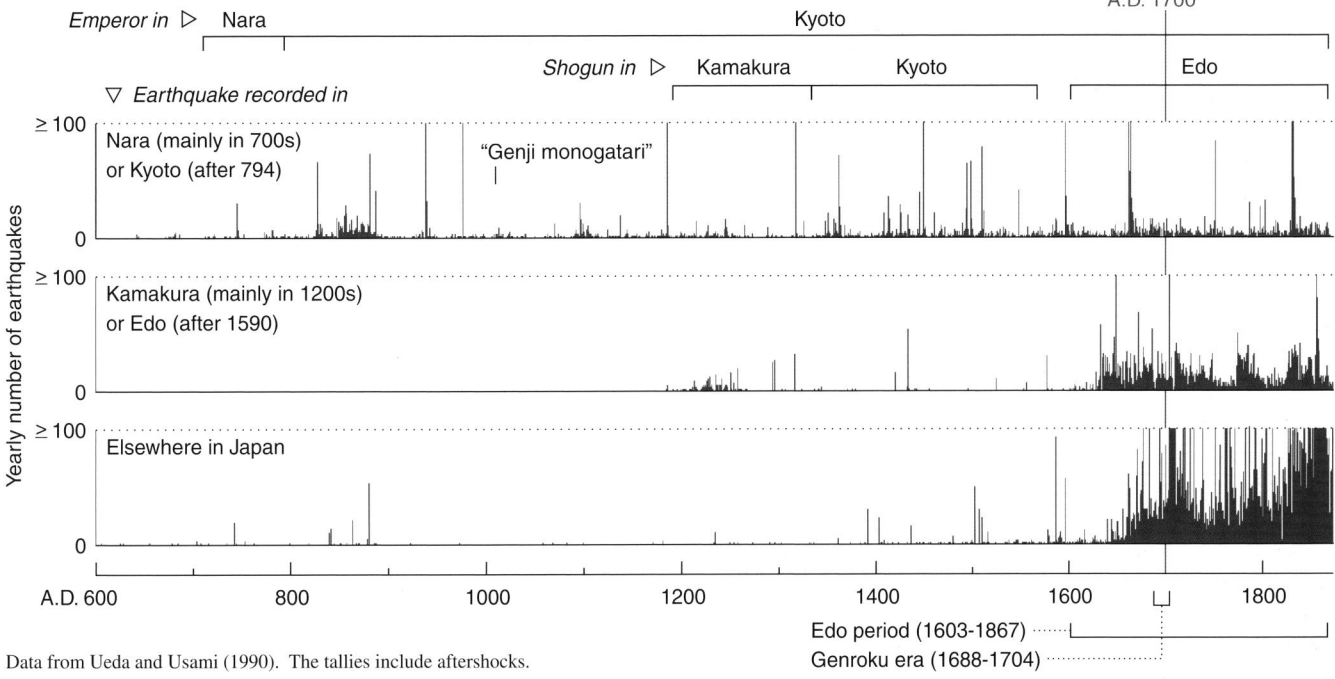

Data from Ueda and Usami (1990). The tallies include aftershocks.

Edo period (1603-1867)
Genroku era (1688-1704)

元 Genroku
禄

THE CULTURAL PEAK of the Edo period is known as Genroku 元禄, the era name for 1688-1704 (p. 42). The Genroku society that kept prodigious records also produced literary innovations, popular titles, and scholarly tomes.

Genroku innovations include haiku—poems of seventeen syllables in three unrhymed lines. These were refined and popularized by Matsuo Bashō (1644-1694). In a posthumous collection he tells how to focus verse so brief: "You should put into words the light in which you see something before it vanishes from your mind."

Bashō's contemporary, Ihara Saikaku (1642-1693), introduced realistic description of the lives of urban merchants and samurai. His novels and collections of short stories include "The life of an amorous man" (1682), "Five amorous women" (1686), "The great mirror of love between men" (1687), "The Japanese family storehouse" (1688), and "Reckonings that carry men through the world" (1692).

The playwright Chikamatsu Monzaemon (1653-1725) popularized puppet theater and wrote dance dramas (*kabuki*) as well. Some of his works treat turmoil in the houses of land barons; others, beginning in 1703, tell of lovers driven to suicide by social obligation and financial difficulty.

Genroku publishers issued thousands of commercial books. A book-dealers' catalog from 1696, in 674 pages, listed 7,800 titles. The books in print included how-to manuals for home use by the young. "Onna chōhōki" (1692) instructed young women; "Otoko chōhōki" (1693),

for young men, provided lessons on calligraphy, Chinese and Japanese poetry, tea ceremony, and letter writing. "Shōbai ōrai" ("Merchants' manual," ca. 1694) exhorted merchants' children to practice writing and arithmetic "from infancy." "Nōgyō zensho" ("Encyclopedia of farming," 1697) advised peasants to "lay up stores of money and grain" as precautions against "the very great risk of dying of starvation in bad years."

An early cookbook, "Ryōri monogatari" ("Story of cooking," 1643), alludes in its title to the classic novel from A.D. 1000, "Genji monogatari" ("The tale of Genji"). "Edo ryōri-shū" ("Collection of Edo cuisine," 1674) filled six volumes.

The shogun throughout the Genroku era, Tokugawa Tsunayoshi (1646-1709), was more scholar than soldier. He is said to have lectured on the "Yiching," an ancient Chinese book of wisdom and divination, no fewer than 240 times between 1693 and 1700.

"Honchō tsugan" ("General history of our State"), completed by 1680, filled 310 volumes. Attributed to the founder of a competing historical project, "Dai-Nihon shi" ("The history of greater Japan"): "In writing one must be true to fact, and the facts must be presented as exhaustively as possible. An excess of detail is preferable to excessive brevity."

Fiction, Kato (1979a, p. 85-112; Bashō quote, p. 102) and Totman (1993, p. 215-220). **Catalog** ("Shojaku mokuroku taizen") and "chōhōki," Shively (1991, p. 720, 731); "Shōbai," Seeley (2000, p. 130); "Nōgyō," Totman (1993, p. 264). **Cookbooks**, Nishiyama (1997, p. 167). **Tsunayoshi**, Bolitho (1976). **Histories**, Tsunoda and others (1964, p. 344-345, 362-364); quotation attributed to Tokugawa Mitsukuni (1628-1700).

Tsunami size 津波の高さ

The 1700 tsunami crested several meters high at the edge of Ōtsuchi Bay.

1700 tsunami

EVIDENCE, HEIGHTS **A** AND **B**

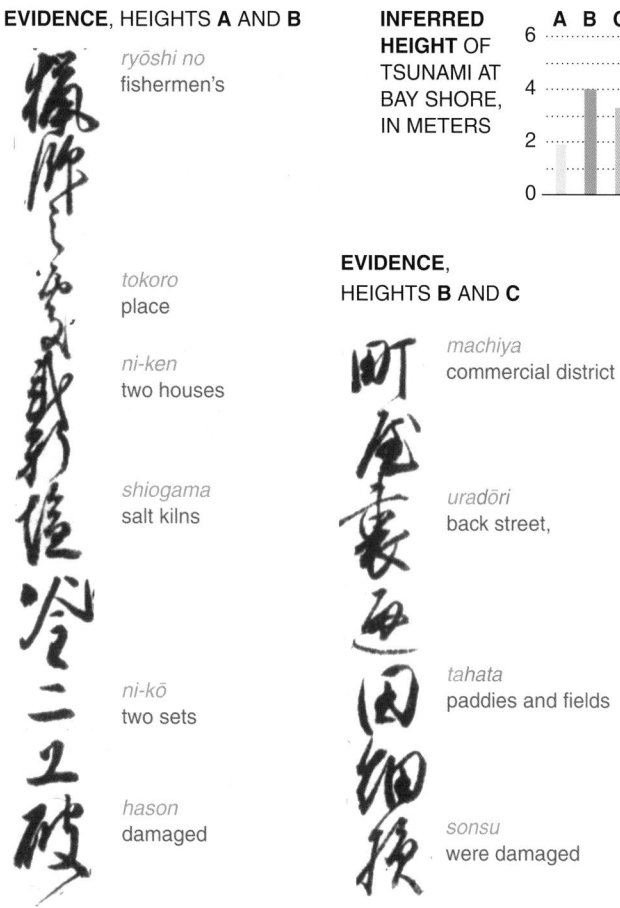

ryōshi no
fishermen's

tokoro
place

ni-ken
two houses

shiogama
salt kilns

ni-kō
two sets

hason
damaged

INFERRED HEIGHT OF TSUNAMI AT BAY SHORE, IN METERS

(bar chart: A B C, scale 0 2 4 6)

EVIDENCE, HEIGHTS **B** AND **C**

machiya
commercial district

uradōri
back street,

tahata
paddies and fields

sonsu
were damaged

WHILE DAMAGING BUILDINGS AND KILNS near the bay shore in front of Ōtsuchi or nearby, the 1700 tsunami approached or exceeded 2 m above tide (**A**).

Maximum tsunami heights of 3-4 m are likely if, like the 1960 Chile tsunami, the 1700 tsunami descended inland to a limit near Ōtsuchi's Edo-period street (first option in **B**). The 1960 tsunami crested 3.6-4.0 m at the bay shore but stopped short of the 2-m contour a few hundreds meters inland (top map, facing page). Independently, maximum tsunami heights of 3-4 m in 1700 can be estimated by assuming that Ōtsuchi has subsided 1-2 m since 1700 (**C** and second option in **B**).

The 1700 tsunami probably crested lower in Ōtsuchi than did the 1677 and 1751 tsunamis, for these flooded houses along the town's main street (p. 59). Unlike estimate **B**, the published height for the 1677 tsunami in Ōtsuchi (2.8 m) lacks correction for an onshore decrease in tsunami height; and unlike estimate **C**, it neglects land subsidence since 1677.

Height **C** from Tsuji and others (1998). Height estimate for 1677 tsunami (p. 59) from Tsuji and Ueda (1995, p. 102).

A Minimum height, inferred from bayside damage

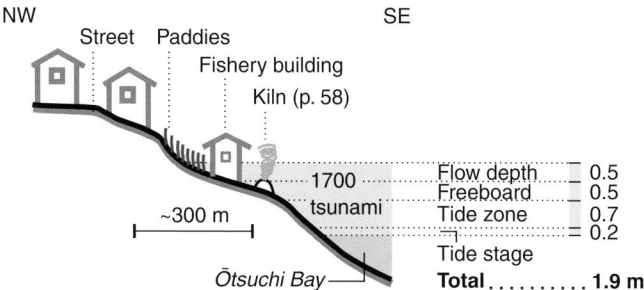

NW   Street   Paddies   SE
Fishery building
Kiln (p. 58)
~300 m
1700 tsunami
Ōtsuchi Bay

Flow depth	0.5
Freeboard	0.5
Tide zone	0.7
Tide stage	0.2
Total	**1.9 m**

ASSUMPTIONS
Flow depth 0.5 m where fishermen's sheds damaged.
Freeboard To have floors above swash during storms—and perhaps with the 1677 tsunami in recent memory—villagers sited these sheds at least 0.5 m above high astronomical tides.
Tide zone Highest astronomical tide in 1700 was 0.7 m above mean sea level, by analogy with modern tides at Kamaishi.
Tide stage 0.2 m above mean sea level; computed for midnight arrival of high water (p. 83).

B More realistic height, inferred either of two ways

Cartoon shows 1960 analogy only

approx. **3-4 m**

Inferred with 1960 analogy The 1700 and 1960 tsunamis both went about the same distance inland. Thus, the 1700 tsunami probably reached heights at the shore similar to those of the 1960 tsunami, in the range 3.6-4.0 m (top map, opposite).
Inferred from subsidence Paddies overtopped by the 1700 tsunami, now at least 1.2 m above mean sea level (brown points on map), may have stood 1.5 m higher in 1700 because of chronic subsidence (estimated in **C**, below).

C Modern land level, adjusted for tectonic subsidence since 1700

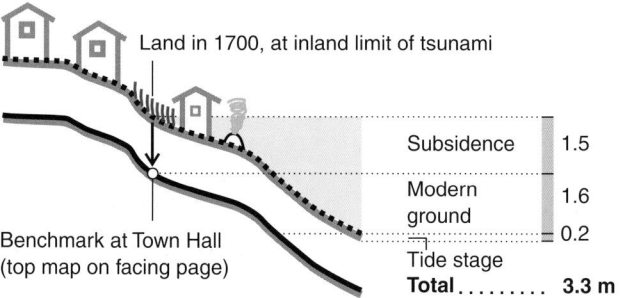

Land in 1700, at inland limit of tsunami

Benchmark at Town Hall (top map on facing page)

Subsidence	1.5
Modern ground	1.6
Tide stage	0.2
Total	**3.3 m**

Subsidence 1.5 m since 1700 (extrapolated tide-gauge trend, right).
Modern ground Maximum height of 1700 tsunami approximated by benchmark 1.6 m above TP.
Tide stage 0.2 m below 1700 mean sea level (p. 83).

THE ORPHAN TSUNAMI OF 1700

1960 tsunami

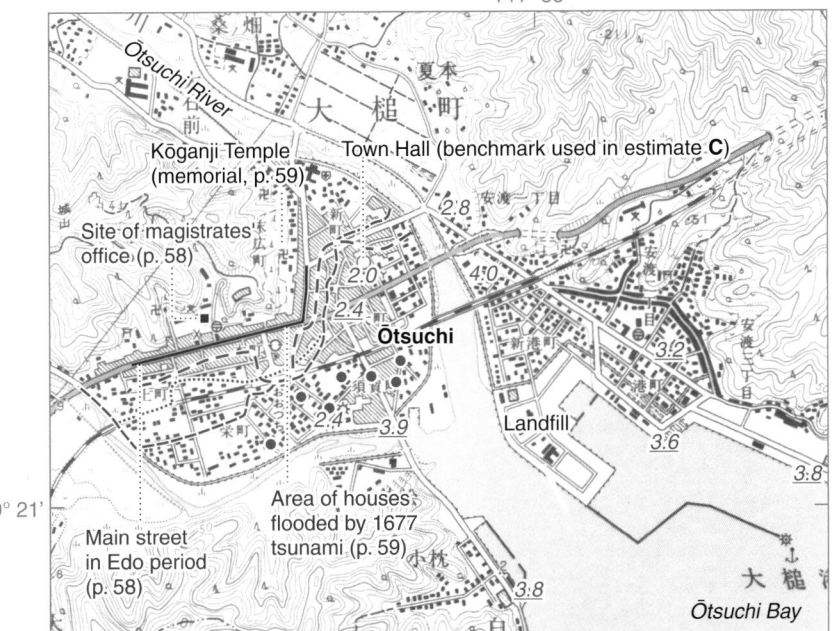

Inland limit of 1960 tsunami   Estimates differ between sources "a" and "b"

.... a

– – – b

Height of 1960 tsunami

3.9 a

2.8 b

Topography mapped in 1975

– – – 2-m contour

● Elevation 1.2 m in former rice paddy

All heights in meters above TP, a datum near mean sea level

Scale 1:25,000

0   1 km

N

a, Omote and Komaki (1961); b, Ōtsuchi-chō Kyōiku I'inkai (1961)

Base map from Kokudo Chiri'in (Geographical Survey Institute), Ōtsuchi 1:25,000 quadrangle, 1996. Contour interval 10 m.

Land-level changes since 1700

THE DESCENDING PACIFIC PLATE dragged land downward along the Japan Trench through the last half of the 20th century. If such subsidence persisted through the last 300 years, northern sites flooded by the 1700 tsunami stood 1.0-1.5 m higher than now—as assumed in the **C** estimates on pages 48, 57, and 64. The assumption is doubtful because (1) 20th-century sea-level rise explains part of the apparent subsidence, (2) stability prevailed at Ayukawa early in the 20th century, and (3) long-term uplift has raised the region's coast in the past 125,000 years.

The coast farther south has a history of cyclic land-level changes related to historical subduction earthquakes on the Nankai Trough (p. 91).

SUBDUCTION ZONES

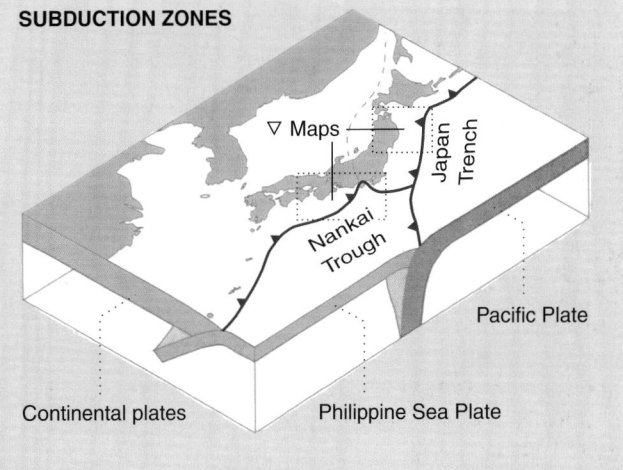

Block diagram modified from Earthquake Research Committee (1998, p. 22). Tide-gauge data from Coastal Movements Data Center (1996); periods of record vary. The mean rates, for 1958-1995, are from Ozawa and others (1997). They include 2 mm/yr of global sea-level rise (Cazenave and Nerem, 2004; Heki, 2004, p. 15). Ota and Omura (1991) infer long-term uplift; Kato (1979b) and Sawai and others (2004) assess historical subsidence.

TIDE-GAUGED TRENDS, LAND LEVEL RELATIVE TO THE SEA

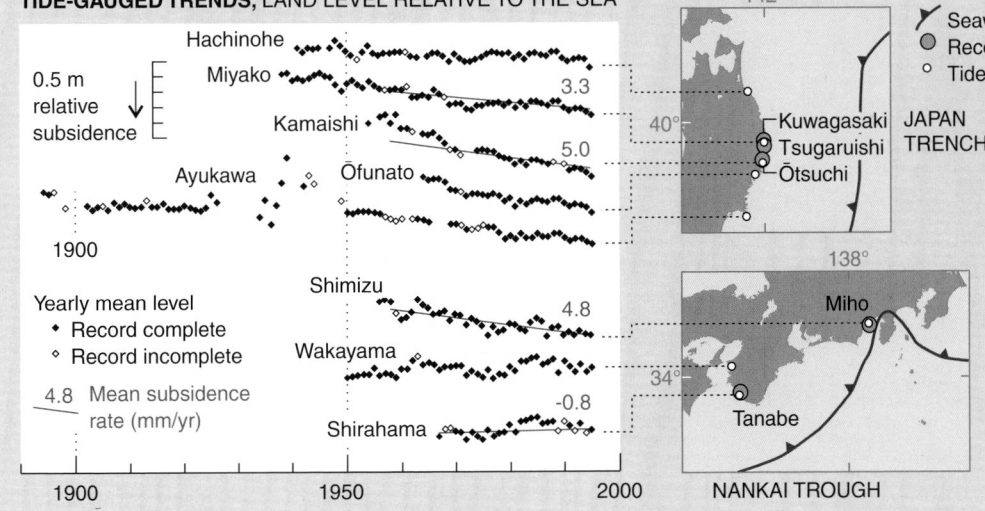

Nakaminato 那珂湊

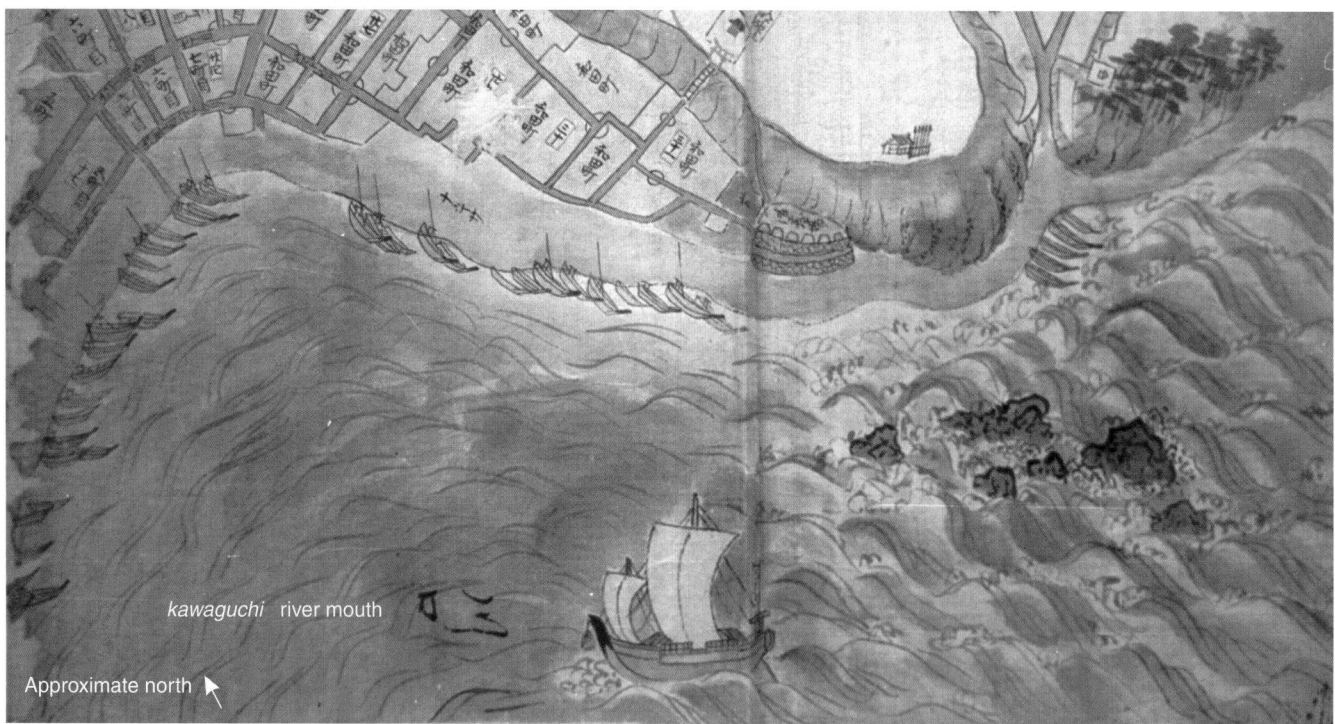

kawaguchi river mouth

Approximate north ➤

At Nakaminato, pictured above in 1842, seagoing boats unloaded Edo-bound cargo that continued to the shogun's capital on inland waterways.

Ōuchi-ke "Go-yōdome," a family's collection of documents on shipwrecks, describes the loss of 28 metric tons of rice from a boat that drifted into rocks at Isohama on January 28, 1700. In 2004 a local historian, Satō Tsugio, compared two writings of the boat captain's name. Ōuchi Yoshikuni (left) represented the family.

THE PICTURE MAP of Nakaminato, above, was made by Watari Kizaemon in Tenpō 13 (1842) and published in Ansei 4 (1857). Map courtesy of the city office of Hitachinaka, Ibaraki Prefecture.

ŌUCHI-KE "GO-YŌDOME," a single volume of 1,390 pages, contains writing from many hands. One hand, probably no earlier than 1735, compiled all the material on wrecks between 1700 and 1735, according to Satō Tsugio, an authority on Mito-han documents. The volume's entire contents, along with reports on 14 other Edo-period shipwrecks, have been printed by Nakaminato Shishi Hensan I'inkai (1993). In this modern volume the 1700 wreck is number 55, pages 81-83.

THE MAPS OPPOSITE are derived from modern sources. The middle map is from Kawana (1984, p. 6, 20) and Kaizuka and others (2000, p. 21). The lowest map is from 1:25,000-scale maps by Kokudo Chiri'in ("Hitachinaka" 1999; "Isohama" 2001) except for the former entrance to Naka River, which is from a 1:200,000-scale map by Rikuchi Sokuryōbu, 1885 (Meiji 18).

THE RICE BOAT went aground at "Hakoiso" (p. 69, column 6). The place name denotes the shore south of the Naka River mouth on a map from 1845 (Tempō 15). The rocks in the photo at right include a group called "Hakoiso" on a fishers' sketch map from Ōarai in 2004.

Main points

High waves on the morning of January 28, 1700, prevented a boat from entering the river-mouth port of Nakaminato. A storm that evening drove the boat to a rocky shore near Isohama village (map, lower right). Lost were all the boat's cargo—28 metric tons of rice—and two of the crew (p. 68-69, 71).

Officials of Mito-han certified the losses in response to a petition (p. 70). The certificate and petition were copied into a family's collection of documents about Edo-period wrecks near Nakaminato (opposite).

The morning high waves probably represent ordinary ocean swells that were opposed at the river mouth by the ebb currents of a long-lasting tsunami (p. 72-75).

Setting

The river mouth at Nakaminato afforded access to inland waterways that conveyed cargo to metropolitan Edo (p. 31, 61). The waterways followed valleys that the sea covered 6,000 years ago. A prehistoric people, the Jomon, fringed this former sea with piles of clam shells (dots, right).

To reach Nakaminato, Edo-period sailors threaded a rocky constriction north of a sand spit (lower map at right, picture map at left). Additional rocks awaited boats that drifted south toward Isohama, a name that means "rocky beach" (photo below).

Nakaminato served as the main port in Mito-han. The rice boat came from another domain, Nakamura-han (upper map). The lost rice belonged to the Nakamura daimyo. Villagers from Isohama towed the wreck for salvage but failed to recover any of the rice.

Documents

The boat captain, two local villagers, and two other men petitioned local officials to certify the accident. The petition and the resulting certificate make up "Ura shōmon no koto" (*ura*, port; *shōmon*, certificate; p. 70-71). A headnote states that 470 bails of rice were lost. Next, a narrative explains the loss in the words of headmen from Isohama village. The certificate concludes with a signed statement by representatives of the senior ministers of Mito-han.

The earliest extant copy appears in a family volume, Ōuchi-ke "Go-yōdome" (*go-yō*, official business; *tome*, records). The volume (opposite) contains documents on 131 shipwrecks near Nakaminato between 1670 and 1832.

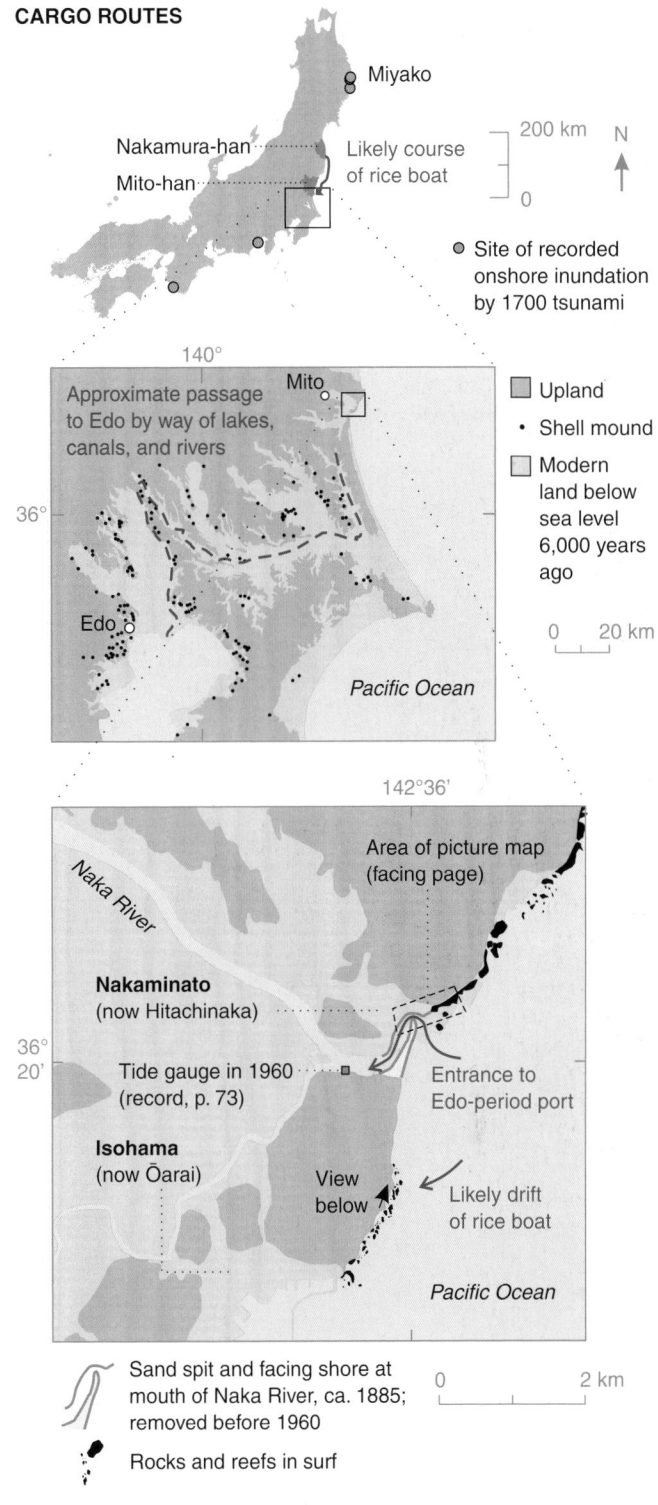

CARGO ROUTES

Miyako

Nakamura-han

Mito-han

Likely course of rice boat

200 km N

0

● Site of recorded onshore inundation by 1700 tsunami

140°

Approximate passage to Edo by way of lakes, canals, and rivers

Mito

36°

Edo

Pacific Ocean

■ Upland

• Shell mound

□ Modern land below sea level 6,000 years ago

0   20 km

142°36'

Naka River

Area of picture map (facing page)

Nakaminato (now Hitachinaka)

36° 20'

Tide gauge in 1960 (record, p. 73)

Isohama (now Ōarai)

Entrance to Edo-period port

View below

Likely drift of rice boat

Pacific Ocean

Sand spit and facing shore at mouth of Naka River, ca. 1885; removed before 1960

Rocks and reefs in surf

0   2 km

Rocks break fair-weather surf near former Isohama.

A BOAT LADEN WITH RICE—470 bales belonging to the daimyo of Nakamura-han—approached the Mito-han port of Nakaminato around 8 a.m. (columns 1-2). High waves held the boat offshore, where the crew cast anchor (3). Still offshore that evening, the crew bailed half the rice during an evening storm (4-5). But the storm broke the anchor lines and drove the boat to a rocky shore near Isohama village (5-6). Two of the crew perished (7).

Afterwards, villagers from Isohama and Nakaminato collected and returned, to surviving crew, articles that had

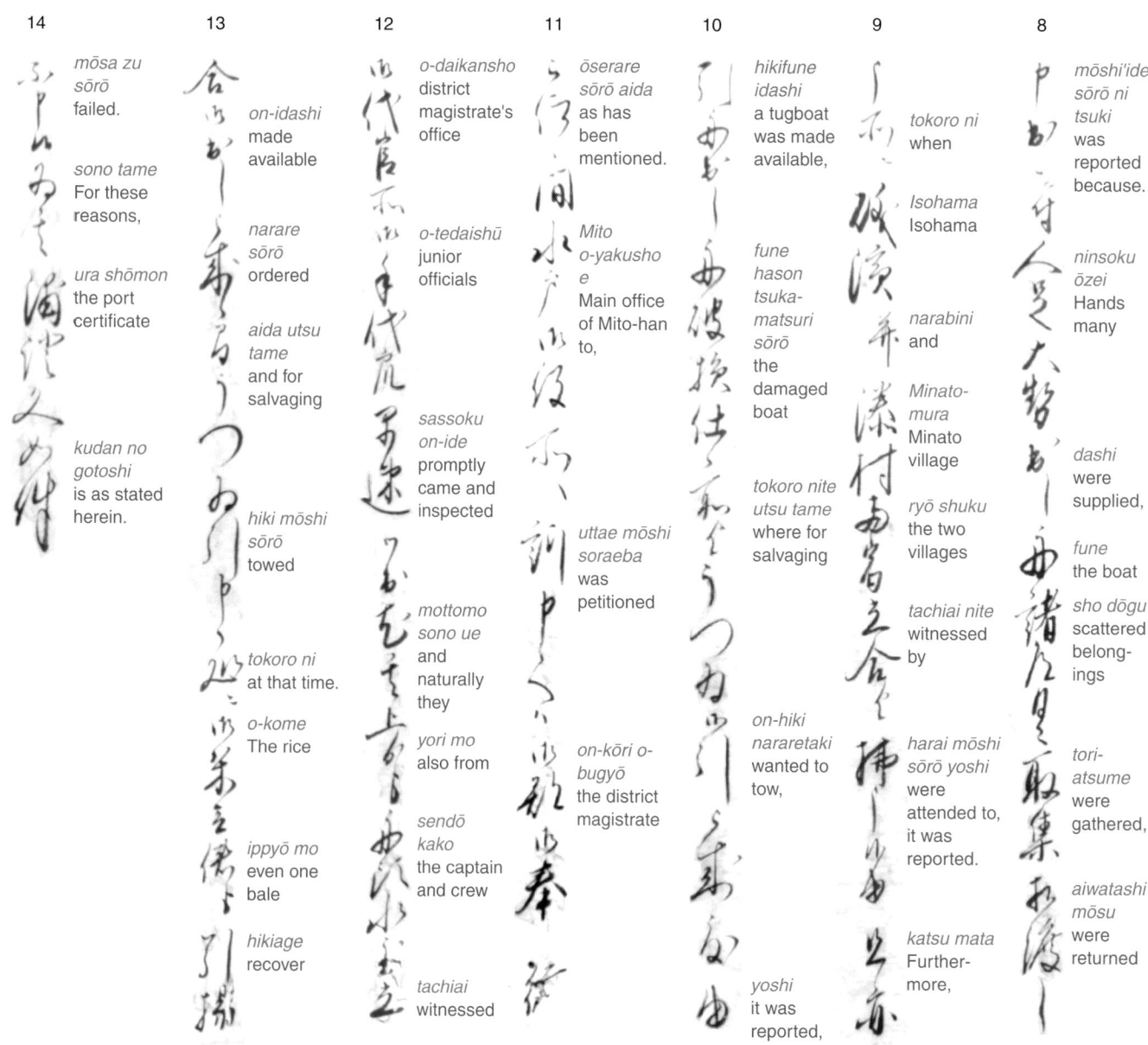

14	13	12	11	10	9	8
mōsa zu sōrō failed.	*on-idashi* made available	*o-daikansho* district magistrate's office	*ōserare sōrō aida* as has been mentioned.	*hikifune idashi* a tugboat was made available,	*tokoro ni* when	*mōshi'ide sōrō ni tsuki* was reported because.
sono tame For these reasons,	*narare sōrō* ordered	*o-tedaishū* junior officials	*Mito o-yakusho e* Main office of Mito-han to,	*fune hason tsuka-matsuri sōrō* the damaged boat	*Isohama* Isohama	*ninsoku ōzei* Hands many
ura shōmon the port certificate	*aida utsu tame* and for salvaging	*sassoku on-ide* promptly came and inspected	*uttae mōshi soraeba* was petitioned	*tokoro nite utsu tame* where for salvaging	*narabini* and	*dashi* were supplied,
kudan no gotoshi is as stated herein.	*hiki mōshi sōrō* towed	*mottomo sono ue* and naturally they	*on-kōri o-bugyō* the district magistrate	*on-hiki nararetaki* wanted to tow,	*Minato-mura* Minato village	*fune* the boat
	tokoro ni at that time.	*yori mo* also from			*ryō shuku* the two villages	*sho dōgu* scattered belongings
	o-kome The rice	*sendō kako* the captain and crew			*tachiai nite* witnessed by	*tori-atsume* were gathered,
	ippyō mo even one bale	*hikiage* recover		*yoshi* it was reported,	*harai mōshi sōrō yoshi* were attended to, it was reported.	*aiwatashi mōsu* were returned
	hikiage recover	*tachiai* witnessed			*katsu mata* Further-more,	

Subject marker *wa* written as *ha* (1, 6); object marker *o* as *wo* (7).

Sound change at word juncture: *dōryō* for *tōryō* (1), *doki* for *toki* (2), *gakari* for *kakari* (3).

12, *kako*—The boat's crew. Printed as *mizu-nushi* (literally, water-owner) in Nakaminato Shishi Hensan I'inkai (1993, p. 82).

Honorific language: *o-* in terms for rice (1, 6, 13), and government offices and titles (11, 12); *on-* in *on-kori* (11), *on-ide* (12), and *on-idashi* (13); *go-* in *go-dōryō* (1). Formal *sōrō* (3, 5, 7, 8, 9, 10, 11, 13, 14), *mōsu* (6). Humble *tsukamatsuri* (3, 5, 10).

6, *Hakoiso*—*hako*, box; *iso*, rock (note, p. 66; photo, p. 67).

7, *yoshi...motte*—Reported by the captain, Kambe'e, through the Isohama villager, Gon'emon. Kambe'e is also the likely source of the second-hand information marked by *yoshi* in column 5.

washed ashore (8-9). Officials of Mito-han, from a district magistrate's office, oversaw a fruitless attempt to find the rice bales (11-14) and certified the accident (14).

The full document, reproduced on the next two pages, contains this narrative as part of a certificate issued to two samurai of unstated affiliation. The narrative's authors, all from Isohama, were a pair of boat headmen, Hei'emon and Rokuza'emon; the village headman, Sakubei; and two village assistant headmen, Heisaku and Jiza'emon.

7	6	5	4	3	2	COLUMN 1
funakata the boat's crew	*fuki-nagasare* drifted.	*minato oki nite* offshore	*dōya* That night	*tsuka-matsuri sōrō tokoro ni* At that time,	*nanuka ni* on the 7th day	*migi wa* At right
ryō-nin two persons	*Hakoiso to* Hakoiso	*uchini* threw over-board,	*itsutsu sugi yori* from the hour of five onwards	*nami* waves	*shussen* the boat departed.	*Sōma Danjō sama* Lord Sōma
aihate were killed,	*mōsu tokoro* a place called	*tsuka-matsuri sōrō yoshi* it was reported.	*daifū'u* strong wind and rain	*takaku* high	*dō* Same	*o-kome* rice
sōrō yoshi it was reported	*nite* at,	*mottomo* However,	*kotoni* especially	*fune* the boat	*kokonuka* 9th day	*go-dōryō* the domain mentioned
sendō Captain	*o-kome wa* the rice	*ikarizuna* anchor ropes	*ō-nami* big waves	*minato e* to the port	*asa no* morning of	*Ukedo-hama nite* at Ukedo harbor
Kambe'e Kambe'e,	*mōsu ni oyoba-zu* not only	*uchikirare* became broken,	*yue* because of	*iri sōrō* enter	*itsutsudoki* at the hour of five	
tōsonjuku this village	*fune* the boat	*fune* the boat	*nangi ni oyobi* it became severe,	*gi* the matter	*tōryō* the domain	
Gon'emon o Gon'emon	*uchi-yaburare* badly damaged;	*tōson chinai e* to this village area	*o-kome* the rice	*makari nara-zu* was unable to,	*Nakano-minato* Nakano-minato	
motte through	*sonoue* moreover,		*kahan* more than half	*funagakari makari ari sōrō* anchored	*omote ni* beside	*tsumitate* loaded,
				tokoro ni then.	*chakusen* the boat reached.	*saru* last

2, *tōryō Nakanominato*—Mito-han's Nakaminato.

4-5, *ō-nami...uchini*—The crew jettisoned cargo in hopes of saving the ship.

5, *tōson chinai*—The boat drifted to the area of Isohama, home of the narrative's authors and of two of the certificate's petitioners, Kichirōemon and Gon'emon (p. 70).

1, *tsumitate*—The rice was loaded onto a boat.

1-2, *saru nanuka ni*—On the most recent 7th day before the headnote date (24th day, 12th month, Genroku 12; p. 70). Similarly, *dō kokonuka* in column 2 means the most recent 9th day.

2 and 4, *itsutsudoki*—About 8 o'clock in the morning (column 2) or evening (4); see page 46.

← NOTES. Column 1, *migi...o-kome*—The rice mentioned previously; itemized, p. 71.

1, *Sōma Danjō sama*—Sōma Masatane served as 5th daimyo of Nakamura-han in 1679-1702.

1, *go-dōryō Ukedo-hama*—Ukedo-hama (literally, Ukedo Beach) was the southernmost of four ports in Nakamura-han (Satō, 1988, p. 167).

Certified loss　浦証文 - 事故の証明

Like a police report on a car crash, a harbor certificate verified the shipwreck.

OBLIGATIONS AWAITED the captain who lost cargo at sea while bound for the Morioka-han port of Kuwagasaki. On arrival he was to inform port officials of the loss. He would then petition them for a port certificate, *ura shōmon*, that could absolve his crew of responsibility while clearing the way for insurance claims.

Similarly in the Mito-han port of Nakaminato, the shipwreck started by the 1700 tsunami resulted in a petition and certificate (below). The petitioners included not just the

captain but also villagers from Nakaminato and Isohama, along with two men we call samurai because they have family names. They addressed their joint petition to officials of Mito-han and Isohama village. In response, village headmen affirmed the accident and han officials, having made an inspection of their own, issued the certificate.

As copied into Ōuchi-ke "Go-yōdome," this ura shōmon contains both the petition and the certificate. Each mentions the "high waves" we ascribe to the 1700 tsunami (p. 73).

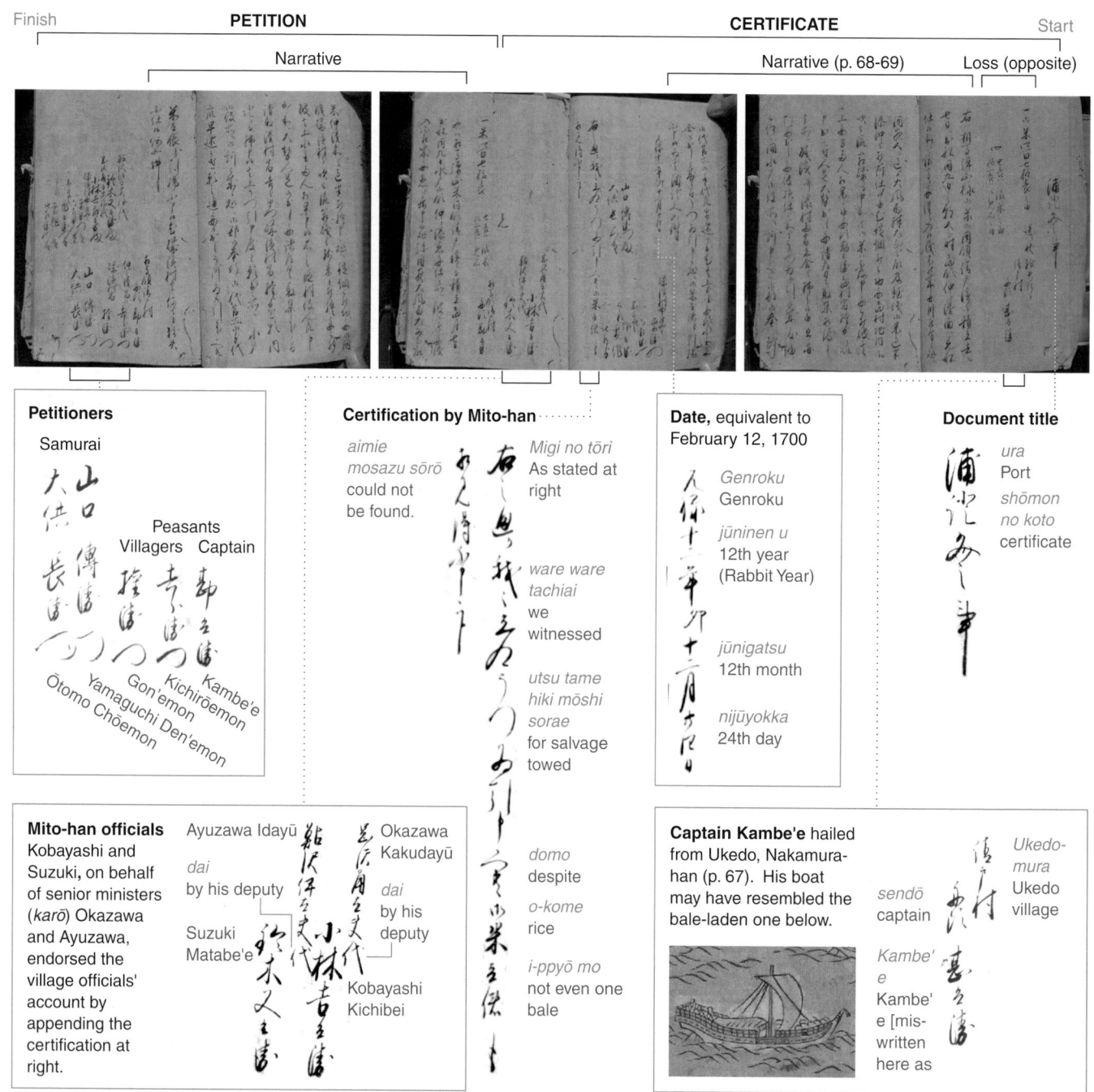

Finish	PETITION		CERTIFICATE	Start
	Narrative		Narrative (p. 68-69)	Loss (opposite)

Petitioners

Samurai

Peasants
Villagers　Captain

Ōtomo Chōemon
Yamaguchi Den'emon
Gon'emon
Kichiroemon
Kambe'e

Mito-han officials
Kobayashi and Suzuki, on behalf of senior ministers (*karō*) Okazawa and Ayuzawa, endorsed the village officials' account by appending the certification at right.

Ayuzawa Idayū
dai by his deputy
Suzuki Matabe'e
Kobayashi Kichibei
Okazawa Kakudayū
dai by his deputy

Certification by Mito-han

aimie mosazu sōrō could not be found.

ware ware tachiai we witnessed

utsu tame hiki mōshi sorae for salvage towed

domo despite

o-kome rice

i-ppyō mo not even one bale

Migi no tōri As stated at right

Date, equivalent to February 12, 1700

Genroku Genroku

jūninen u 12th year (Rabbit Year)

jūnigatsu 12th month

nijūyokka 24th day

Captain Kambe'e hailed from Ukedo, Nakamura-han (p. 67). His boat may have resembled the bale-laden one below.

sendō captain

Kambe'e Kambe'e [mis-written here as

Document title

ura Port

shōmon no koto certificate

Ukedo-mura Ukedo village

The certificate begins by itemizing the loss of 470 bales of rice (right). Those bales probably looked like the ones that burly men fill, cinch, lift, and carry in the Hokusai sketches below. Each bale, with a volume of one *hyō* (*i-ppyō*), probably weighed close to 60 kilograms (130 pounds).

Two and a half bales made up one *koku*. A unit of volume, the koku measured such quantities as the capacity of freighters. But it also measured wealth and status—the amount of rice granted annually to a samurai (the 50-koku stipend of the former merchant, Moriai Chūzaemon, p. 53), and the officially expected agricultural yields that ranked daimyo domains (examples, below right).

uchi
including

jūsan-byō wa
13 bales

nana-hyō wa
7 bales

go-mengoku
tax-exempt rice

rōmai no
rice to be consumed
[by the crew]

no yoshi
it was reported

no yoshi
it was
reported

o-kome
Rice

*yonhyaku
nanaji-ppyō no*
470 bales of

ppyō, byō—same
meaning as *hyō*

"Men baling rice" From a book of sketches by Katsushika Hokusai (1760-1849).

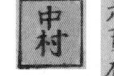

REPUTED YIELDS
OF DAIMYO DOMAINS

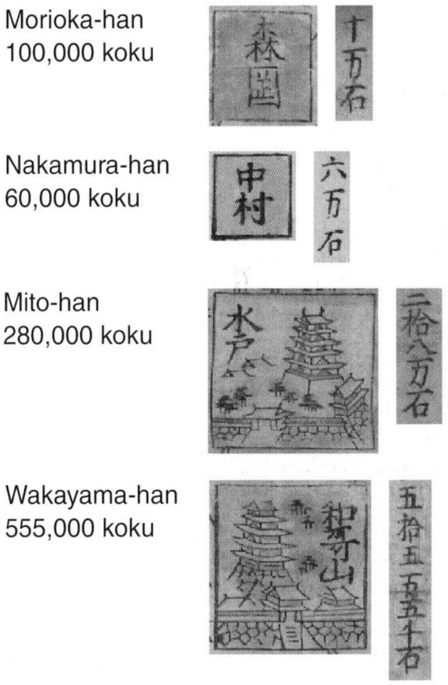

Morioka-han
100,000 koku

Nakamura-han
60,000 koku

Mito-han
280,000 koku

Wakayama-han
555,000 koku

Morioka-han — Kuwagasaki
Nakamura-han
Mito-han — **Nakaminato**
○ Known site of
1700 tsunami
Wakayama-han
N
500 km

ON CERTIFICATION OF SHIPWRECKS, see Miyako-shi Kyōiku I'inkai (1981, p. 498-519) and wrecks 49 and 56 in Nakaminato Shishi Hensan I'inkai (1993).

BOAT AND YIELDS from "Nihon kaisan chōriku zu," 1694 (p. 30-31), courtesy of the East Asian Library, University of California, Berkeley.

"MEN BALING RICE" from "Hokusai manga" ("The sketches of Hokusai"), v. 3, page 6r. Woodblock-printed album published in 1850 by Eirakuya Tōshirō and Kadomaruya Jinsuke. Courtesy of The Art Institute of Chicago, image 761.952.

Fair-weather waves 好天下の高波

The 1700 tsunami in Japan began without a storm but may have continued into one.

THE SUN WAS SHINING from Morioka to Wakayama the day before the 1700 tsunami approached Japan (the 7th day, below). On the 8th day, as the tsunami crossed the Pacific (p. 74-75), skies remained fair over Morioka and Wakayama while snow fell in Edo. Rain or snow fell widely on the 9th day, but mainly in the evening and not at Morioka or Nikkō.

Most of these weather observations come from diaries.

Some are official journals—from castle towns, a shrine, a temple, and Edo mansions (p. 61). Others were kept by court aristocrats in the imperial capital, Kyoto.

Among narratives of the 1700 tsunami, only the Nakaminato rice-boat story mentions weather—a storm that arrived 12 hours after the crew first encountered "high waves" as they tried to enter port.

Weather observations

7TH DAY
Kuwagasaki
Morioka (a)
HONSHU
Edo (c-e)
Kyoto (g-j)
Ōtsuchi
Wakayama (k)
Nakaminato
Miho
PACIFIC OCEAN

Sun Clouds Rain Snow

8TH DAY
Morioka-han
Murakami-han
Nagoya (f)
Wakayama-han
c d e
g h i j
Tsushima-han

9TH DAY
Nikkō (b) clouds late
Rain late Storm in evening
g h i j

0   500 km   N

TIMELINE FOR 1700 TSUNAMI
12th month of Genroku 12 (p. 42-43)

Begins along west coast of North America	
Noticed in Kuwagasaki and Ōtsuchi	
Unreported but likely at all sites	
Noticed off Nakaminato and in Miho	
Wanes in Miho	
Continues into storm off Nakaminato?	
Storm arrives	

day — night

7th day — 8th day — 9th day

Weather observers

LOCATION		DIARY AND WRITER
a	Morioka	**"Morioka-han zassho"** Administrators of Morioka-han (p. 44, 60).
b	Nikkō	**"Shake gobansho nikki"** Officials of shrine for the grave of Tokugawa Ieyasu (shogun, p. 41).
c	Edo	**"Gokokuji nikki"** Buddhist monks.
d	Edo	**"Sakakibara-ke Edo hantei nikki"** Officials at an Edo mansion of the Sakakibara family, which then ruled Murakami-han. Diary started 1651, continued to 1866; 553 volumes. For map of Edo mansions of daimyo like the Sakakibara, see pages 61 and 106.
e	Edo	**"Tsushima-han Edo hantei mainikki"** Official diary of an Edo mansion of the Sō family, daimyo of Tsushima-han.
f	Nagoya	**"Ōmu rōchū ki"** Asahi Bunzaemon Shigeaki, floor-mat manager (*tatami bugyō*) of Nagoya castle. The castle was headquarters of one of the three main branches of the Tokugawa family. As the caged parrot (*ōmu rōchū*) in the book's title, Asahi says he wrote exactly what he heard.
g	Kyoto	**"Kinsumi-kyō ki"** Shigenoi Kinsumi, court aristocrat and scholar.

LOCATION		DIARY AND WRITER
h	Kyoto	**"Kinmichi ki"** Ōgimachi Kinmichi, court aristocrat and Shinto scholar.
i	Kyoto	**"Tokudaiji hinami"** Tokudaiji Kōzen, court aristocrat.
j	Kyoto	**"Sadamoto-kyō ki"** Nonomiya Sadamoto, court aristocrat and scholar.
k	Wakayama	**"Miura-ke nikki"** Miura-family head serving as a karō (senior minister; p. 44) of Wakayama-han.

WEATHER OBSERVATIONS are lacking from Nikkō on the 7th and 8th days, and from Nagoya on the 7th day. Observations differ in Edo on the 8th day, in Kyoto on the 8th and 9th days. All were first compiled in Tsuji and others (1998, p. 8), where Ueda mislocated observation g in Ise (80 km south of Nagoya).
e — All Korean trade sanctioned by the Tokugawa shogunate passed through Tsushima-han (Totman, 1993, p. 76-77).
f — Printed by Nagoya-shi Kyōiku I'inkai (1965-1969). Nagoya castle contained 50,000 m^3 of lumber and stood until World War II (Totman, 1989, p. 62; Naito and Hozumi, 2003, p. 52, 63). Sketch from "Nihon kaisan chōriku zu," 1694 (p. 30), courtesy of East Asian Library, University of California, Berkeley.
JAPAN'S WINTER STORMS "cause ship disasters as well as damage along the coast due to wind waves" (Arakawa and Taga, 1969, p. 128). They are not typhoons, which instead hit Japan in summer and fall (p. 83).

Waves raised by an opposing current

The morning "high waves" that held the rice boat offshore probably originated as incoming ocean swells that met river-mouth backwash of a long-lasting tsunami.

Several accounts refer to the 1700 tsunami as a tide (p. 40). The Miho headman, for instance, reports that the water came in "something like a very high tide" about seven times between dawn and about 10 a.m. ("the hour of four"). The headman further notes that the water drained "with the speed of a big river" (p. 79, columns 3-4).

Such tide-like currents impressed eyewitnesses to the 1960 tsunami at Nakaminato. They estimated incoming velocities at 7 knots (about 3.5 meters per second) and described the outflow as even faster.

Strong ebb currents heighten incoming ocean waves on river-mouth bars. An Oregon boating manual warns, "If you are trapped outside a rough bar on an ebb tide, it is wise to lay to and wait" until a rising tide produces an inflowing current.

The 1700 tsunami probably produced strong ebb currents that heightened waves at 8 a.m. off Nakaminato. Such currents should not have resulted from the astronomical tide, which was rising at that hour from Kuwagasaki to Tanabe (p. 83). But the tsunami, at Miho, was then producing intermittent, swift outflow. Similar outflow from the port of Nakaminato probably raised the "high waves" that eventually led to the rice boat's demise.

The tsunami likely continued raising river-mouth waves through the morning and perhaps into the early evening. It disturbed seas at Miho until noon (p. 79, columns 4-5). Together with the coming storm it may explain why the rice boat stayed off Nakaminato throughout the day.

An outsize tsunami can go on for 24 hours or more. The 1960 tsunami lasted that long (marigrams below and p. 46). Similarly in a computer model, the 1700 tsunami disturbs the Pacific Ocean for an entire day (next two pages).

Duration of tsunami wave trains in 1700 and 1960

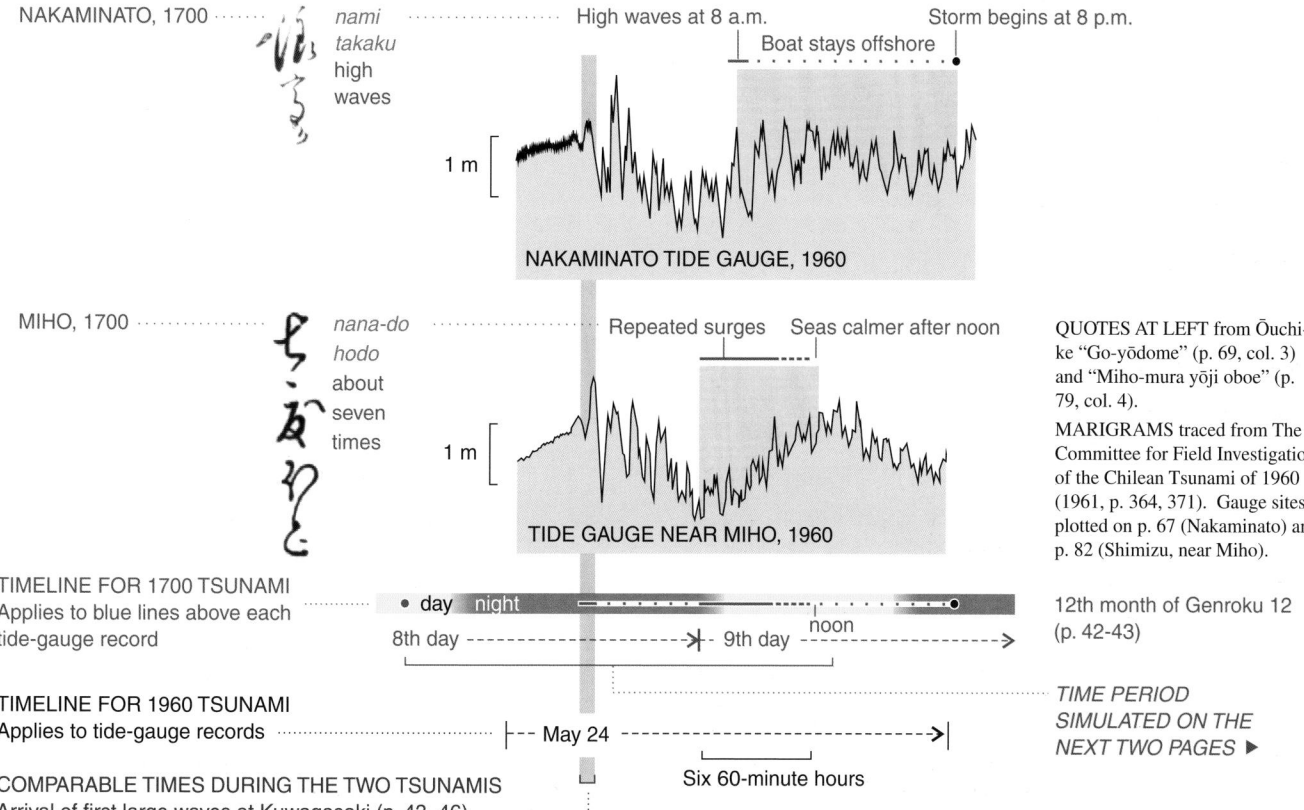

NAKAMINATO, 1700 · · · · · *nami takaku* high waves · · · · · High waves at 8 a.m. · · · · · Storm begins at 8 p.m.

Boat stays offshore

1 m

NAKAMINATO TIDE GAUGE, 1960

MIHO, 1700 · · · · · *nana-do hodo* about seven times · · · · · Repeated surges · · · · · Seas calmer after noon

1 m

TIDE GAUGE NEAR MIHO, 1960

QUOTES AT LEFT from Ōuchi-ke "Go-yōdome" (p. 69, col. 3) and "Miho-mura yōji oboe" (p. 79, col. 4).

MARIGRAMS traced from The Committee for Field Investigation of the Chilean Tsunami of 1960 (1961, p. 364, 371). Gauge sites plotted on p. 67 (Nakaminato) and p. 82 (Shimizu, near Miho).

TIMELINE FOR 1700 TSUNAMI
Applies to blue lines above each tide-gauge record

• day night

8th day - - - - - - - - - - - → 9th day noon - - - - - - - →

12th month of Genroku 12 (p. 42-43)

TIMELINE FOR 1960 TSUNAMI
Applies to tide-gauge records · · · · · — May 24 - - - - - - - - - - - →

TIME PERIOD SIMULATED ON THE NEXT TWO PAGES ▶

Six 60-minute hours

COMPARABLE TIMES DURING THE TWO TSUNAMIS
Arrival of first large waves at Kuwagasaki (p. 43, 46) · · · · · · ·

CURRENTS. Toba and Taka (1961, p. 309) summarize eyewitness accounts of currents at Nakaminato during the 1960 tsunami. The Oregon boating manual offers little comfort for the latter-day Kambei who becomes "trapped outside a rough bar with a southwester developing 40-knot or better winds... If possible, run to another port with more favorable bar conditions" (Oregon Sea Grant and Oregon State Marine Board, 1999, p. 8-9). On the bar off the mouth of the Columbia River (river location, p. 22), wave height oscillates at tidal periods and peaks during ebb currents, which at this bar commonly exceed 2 meters per second. During a five-day series of measurements, ebb currents raised incoming waves of 3 meters to heights as great as 7 meters (González, 1984).

WAVE PERIODS. During a typical wave of the 1960 tsunami at Nakaminato (above), the crest-to-trough fall in water level amounted to 1 meter and took an hour or two. In contrast, the port's astronomical tides change water levels by no more than 1.8 m in six hours, as judged from extreme tides 20 km south of Nakaminato (at Kashima; Maritime Safety Agency, 1998).

Simulated waves 津波のシミュレーション

In a computer model, a long-lasting 1700 tsunami engulfs the Pacific.

TIME IN JAPAN
(~120-minute hours; p. 43)

TIME SINCE EARTHQUAKE
(60-minute hours)

→ 0

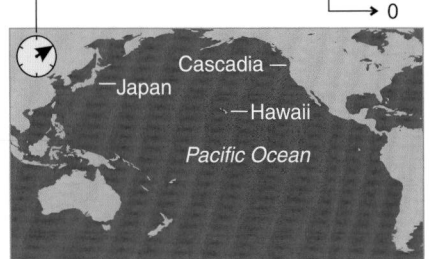

Hour of eight, afternoon of 8th day, 12th month, 12th year, Genroku era
Cascadia earthquake, ~9 p.m., Jan. 26, 1700

1

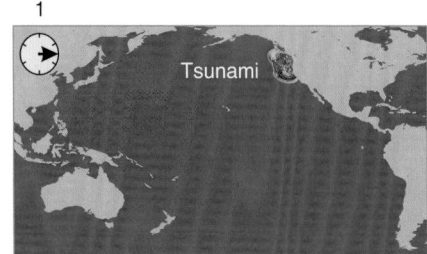

Tsunami already striking Pacific coast at Cascadia (resulting deposits, p. 18, 20)

2

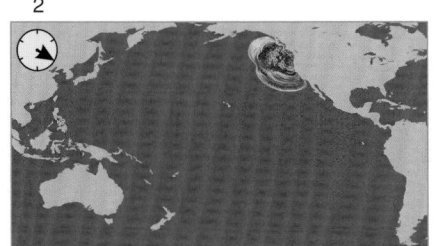

Hour of seven, afternoon, 8th day of 12th month

6

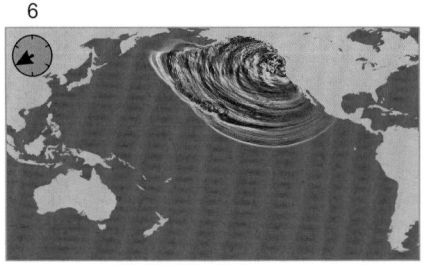

Hour of five, evening, 8th day of 12th month

7

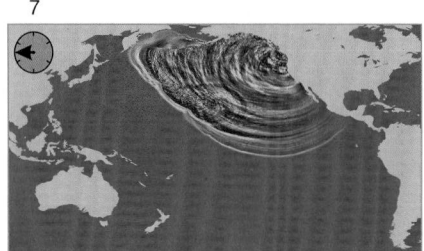

8

Hour of four, night, 8th day of 12th month

12

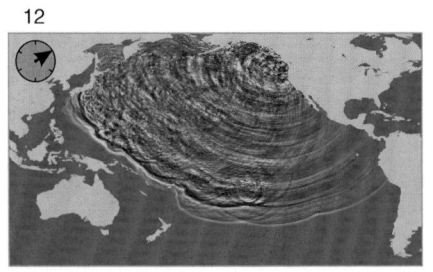

Hour of eight, night, 8th day of 12th month
At Cascadia, 9 a.m. of January 27

13

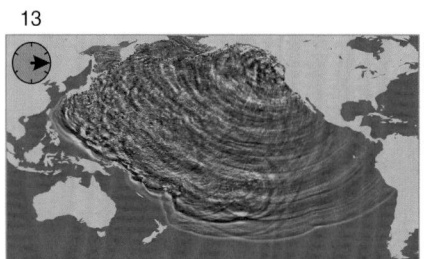

14

Hour of seven, night, 8th day of 12th month

18

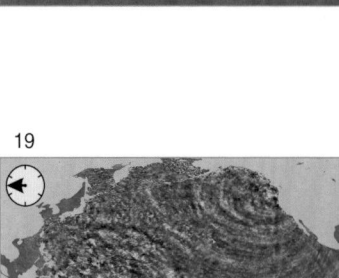

Hour of five, morning, 9th day of 12th month
"High waves" off Nakaminato (p. 69, col. 2-3)

19

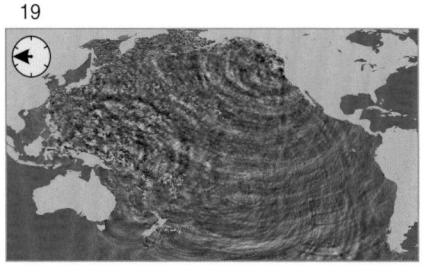

20

Hour of four; morning, 9th day of 12th month
About seven daylight waves noted by then at Miho (p. 79, col. 4)

IN THE DEEP OCEAN a tsunami's waves have little height but great crest-to-crest length. As they enter shallow water the waves slow down and stack up. In the model above, the 1700 tsunami rarely rises more than 0.5 m as it crosses the Pacific but builds against Japanese shores to heights as great as 5 m (p. 99).

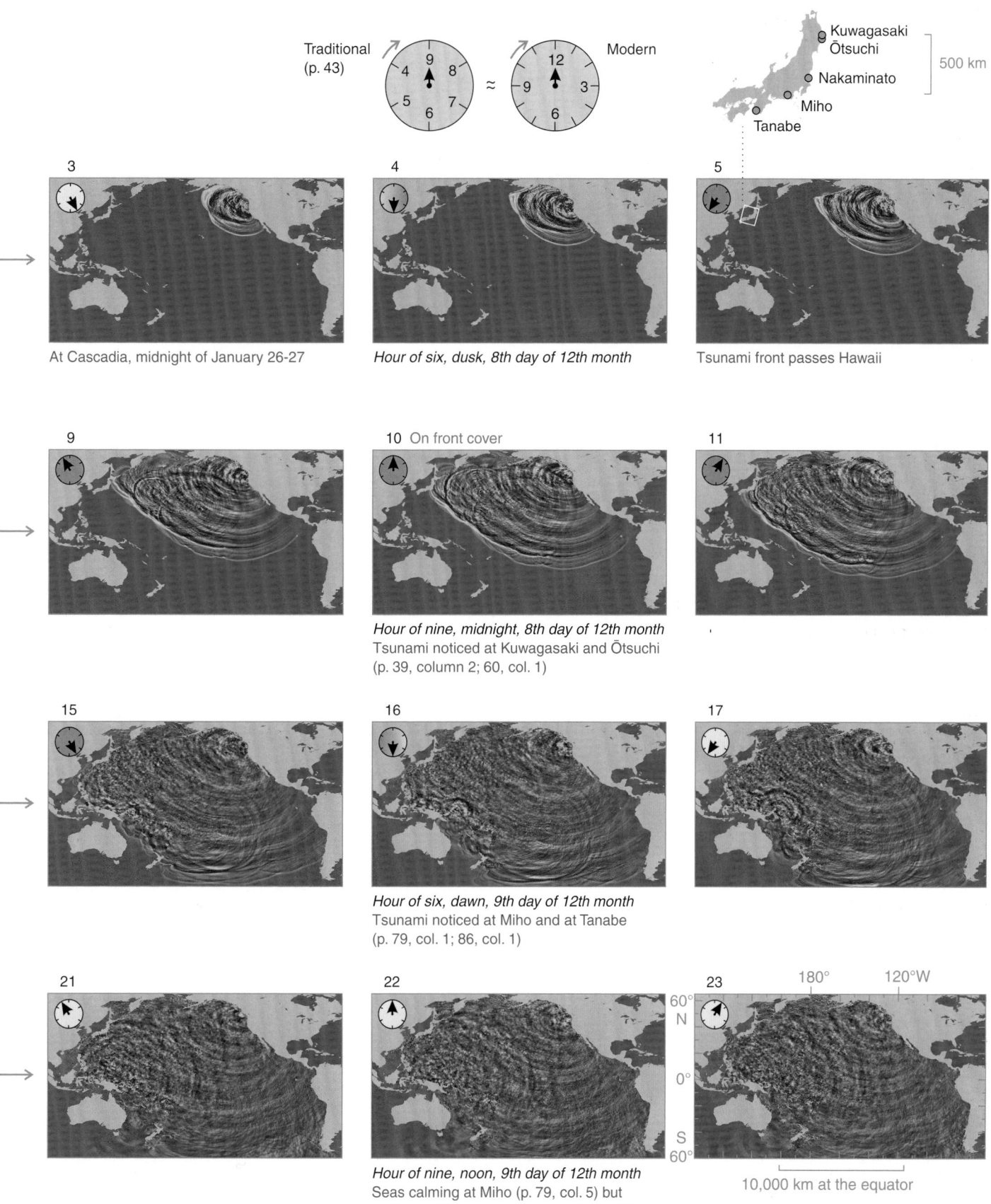

Traditional (p. 43) ≈ Modern

3

At Cascadia, midnight of January 26-27

4

Hour of six, dusk, 8th day of 12th month

5

Tsunami front passes Hawaii

9

10 On front cover

Hour of nine, midnight, 8th day of 12th month
Tsunami noticed at Kuwagasaki and Ōtsuchi
(p. 39, column 2; 60, col. 1)

11

15

16

Hour of six, dawn, 9th day of 12th month
Tsunami noticed at Miho and at Tanabe
(p. 79, col. 1; 86, col. 1)

17

21

22

Hour of nine, noon, 9th day of 12th month
Seas calming at Miho (p. 79, col. 5) but
probably still rough at Nakaminato (p. 73)

23

180°   120°W

60° N   0°   S 60°

10,000 km at the equator

THE MODEL depicts the tsunami from a Cascadia earthquake of magnitude 9.0 with a fault rupture 1,100 km long (p. 98-99; Satake and others, 2003).

ANIMATED VERSION of the model:
ftp://www.agu.org/apend/jb/2003JB002521/2003JB002521-animation.gif

Miho 三保

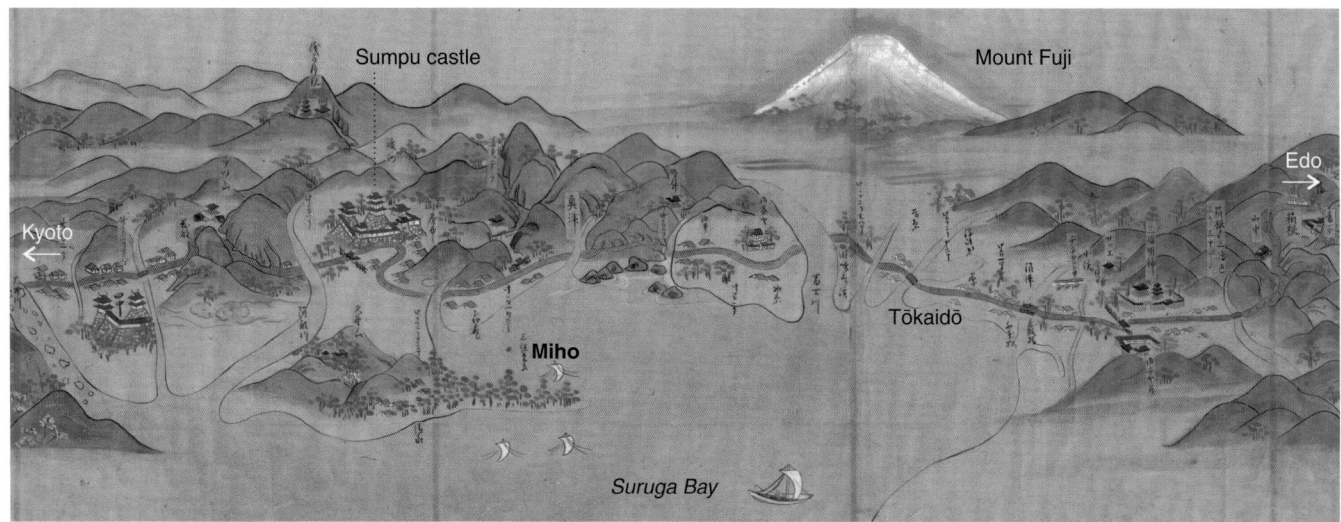

Sumpu castle   Mount Fuji

Kyoto ←

Edo →

Miho

Tōkaidō

Suruga Bay

←— West and southwest

1687

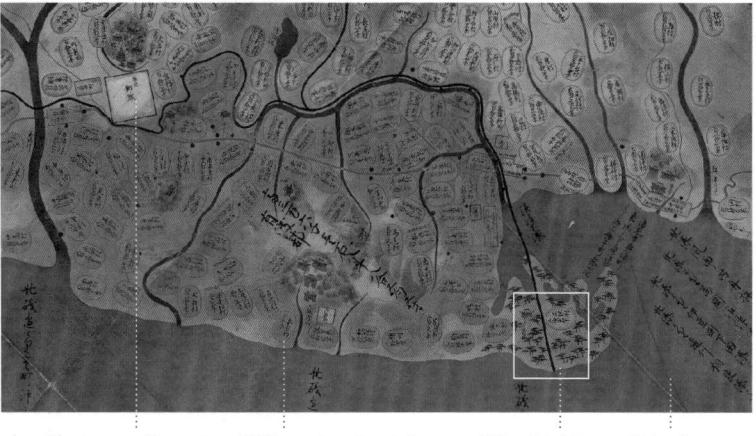

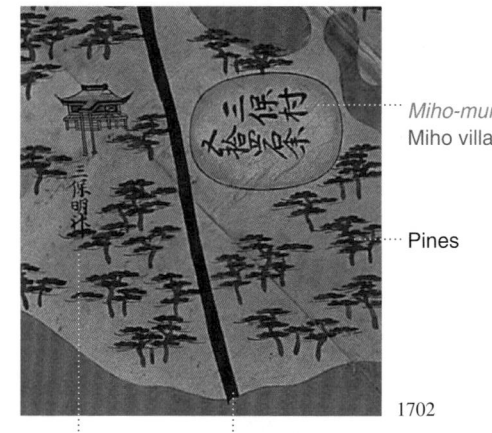

Miho-mura
Miho village

Pines

1702

| Sumpu castle (p. 41) | Village (each oval) | Miho (right) | Shipping route (p. 33) | *Miho myōjin* Miho shrine | Boundary between two of the seven counties in Suruga province (p. 31) |

Miho village, on a sheltered shore of a pine-covered sand spit, was home to some 300 peasants in 1700. To their headman, the tsunami of that year resembled a series of brief high tides. He noted that the waves lacked a parent earthquake felt in Miho or nearby. His inquisitive eyewitness account survives in "Miho-mura yōji oboe," a selection of headmen's records probably compiled in the early 19th century. At right, the document and its possessor in 1999, hotel owner Endō Kunio.

THE OBLIQUE VIEW at top comes from "Tōkaidō narabini saigoku dōchū ezu," a picture map dated Jōkyō 4 (1687). The full map unfolds to a length of 8.7 meters. In the reduced image above, which represents one-tenth of that span, Miho's pine-covered spit commands a view of Mount Fuji. Not far from the village, the Tōkaidō—the famed Eastern Sea Road between Edo and Kyoto (p. 30-31; Naito and Hozumi, 2003, p. 32-33)—winds past Sumpu castle, where a diarist for Tokugawa Ieyasu wrote "tsunami" in its modern characters (p. 41). Courtesy of East Asian Library, University of California, Berkeley.

THE PLAN VIEWS are details from a map of Suruga province, "Suruga no kuni," dated Genroku 15 (1702). The entire map spans 4.5 m by 3.9 m; the detail at left,

70 cm by 35 cm. The closeup at right shows Miho shrine labeled *myōjin*, illuminating god. From the collection of Ashida Koreto (1877-1960); map 54-35 of Ashida Bunko Hensan I'inkai (2004, p. 226). Courtesy of the library of Meiji University, Tokyo.

"MIHO-MURA YŌJI OBOE" Endō Kunio, 85 years old in 1999, inherited "Oboe" from his grandfather. Endō Shōji edited a pair of volumes on "Oboe" that reproduce all its cursive entries, print them in old characters, transcribe these into modern characters, and place the accounts in historical context (Endō and Nagasawa, 1989; Endō and others, 1990). An "Oboe" entry from Hōei 3 (1706) gives the village's population as 328 and its number of houses as 54.

Main points

While waves held the rice boat off Nakaminato (p. 73), the sea at Miho rose and fell repeatedly, like a swift series of tides (p. 78-80).

Wary of flooding, Miho's headman advised villagers to flee (p. 46; 79, columns 6-7).

The lack of an associated earthquake puzzled the headman, who expected that an earthquake would precede a tsunami (p. 40; 54; 78, columns 12-14).

Like the 1960 tsunami, the 1700 tsunami at Miho was probably under 2 m in height and caused less flooding than did a storm of its era. The largest wave of the 1960 tsunami at Miho probably rode a higher tide than did the 1700 tsunami (p. 82-83).

Documents

Compiled in the 19th century from records of the village's headmen, "Miho-mura yōji oboe" contains 71 entries on a wide range of topics. Three-quarters of these date from 1694 to 1730.

The 1700 tsunami appears twice in "Oboe"—as the main event in the account on the next two pages, and as a flashback in the entry below, which begins with an earthquake that devastated Edo in December 1703. The tsunami from this earthquake reminded the headman of Rabbit Year waves—doubtless those of Genroku 12 (p. 42). In both cases, the likely writer was either Gorōemon or Chūemon, the village's headmen in 1698.

1703 EARTHQUAKE AND TSUNAMI	1700 TSUNAMI

Detail at right ← ←|

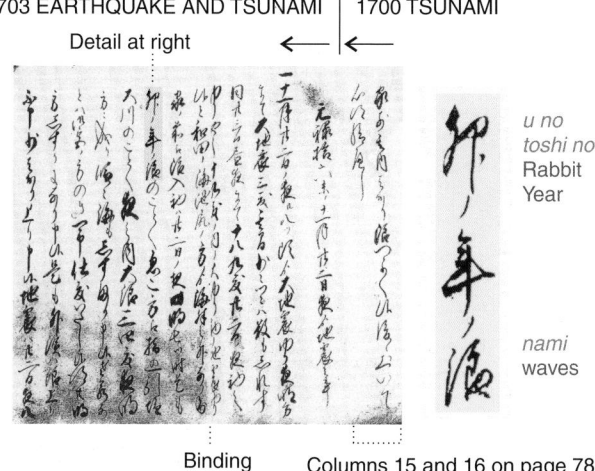

u no toshi no Rabbit Year

nami waves

Binding

Columns 15 and 16 on page 78

"MIHO-MURA YŌJI OBOE" also recounts: **1697**, Fire destroys 23 homes—40 percent of the village's houses—when wind fans flames from the home of Zen'emon. Later, villagers dry a big catch of sardines and the shogunate orders a new map of the area (probably the 1702 map, opposite). **1698**, Gorōemon and Chūemon, as *shōya* (headmen) of Miho, sign a land survey that assesses Miho's cropland at 54.2 koku (expected yield: 9,756 liters of rice, or its equivalent). **1699**, Destructive typhoon (p. 83). **1713**, Villagers build a salt-water fish pond in front of the headman's house. They capture a wild pig after it attacks children. **1715**, Inflation raises the price of rice; villagers die of malnutrition. **1719**, The shogunate, through a merchant in Edo, orders fresh fish from Miho for a 500-person delegation from Korea. **1722**, A wet summer hurts salt production. **1729**, An elephant passes through a nearby village while on its way to Edo. Seven years old, it has a height of 6 *shaku* 5 *sun* (2 m) and a length of 1 *jō* (3 m). Its tail extends 3 shaku 3 sun (1 m), its tusks 1 shaku 3 sun (0.4 m). Its ears are shaped like ginko leaves, its eyes like leaves of bamboo. Its daily diet includes 100 tangerines, 6 *sho* (11 liters, or 3 gallons) of cooked rice, and 9 sho of sake.

PLATE-BOUNDARY EARTHQUAKES

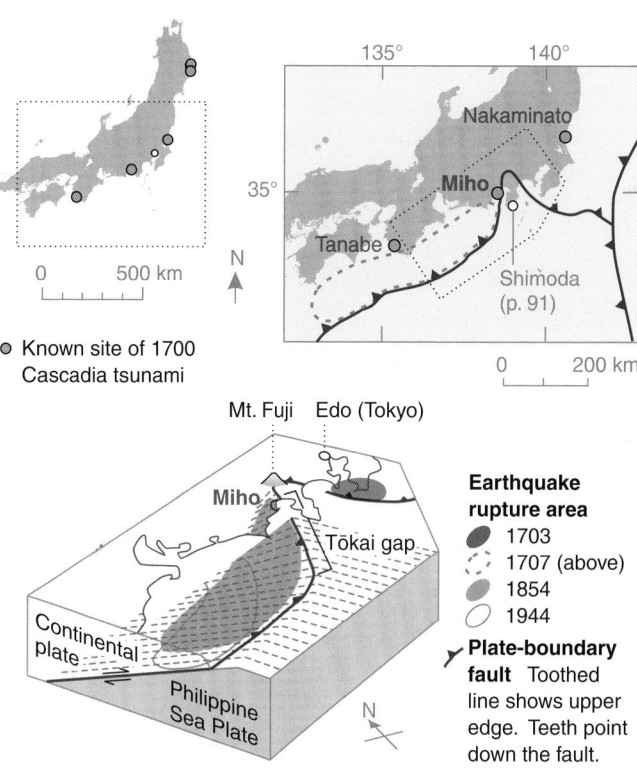

● Known site of 1700 Cascadia tsunami

Earthquake rupture area
- 1703
- 1707 (above)
- 1854
- 1944

⊢ **Plate-boundary fault** Toothed line shows upper edge. Teeth point down the fault.

Tectonic setting and local tsunamis

Barely ten kilometers of continental crust separates Miho from the giant fault on which the Philippine Sea tectonic plate descends. Beneath Miho, this fault ruptured during an earthquake in 1854, and perhaps in 1707, but not in 1944. Efforts to predict earthquakes in Japan focus largely on the patch that failed to break in 1944—the Tōkai gap.

The earthquakes of 1707 and 1854 produced tsunamis as much as 5 m high in Miho. Lesser tsunamis in Miho resulted from nearby earthquakes in 1605, 1703, and 1944.

NOTABLE TSUNAMIS IN MIHO SINCE 1600

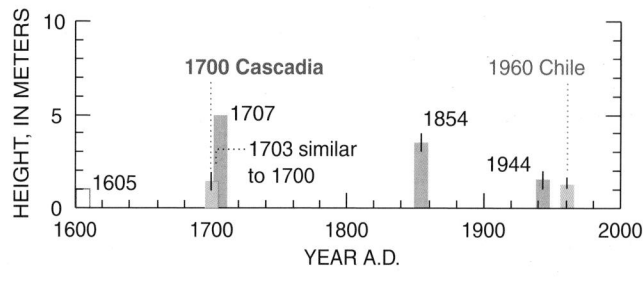

- ■ Tsunami generated by nearby earthquake
- ▨ Tsunami from distant source
- □ 1605 tsunami likely at Miho but height there unknown
- | Range of estimates or measurements

ON HISTORICAL EARTHQUAKES near Miho and forecasts for the Tōkai gap, see Ishibashi (1981). We traced the block diagram from his page 312.

THE TSUNAMI HEIGHTS are from Watanabe (1998, p. 71, 80, 93, and 133) for 1605 and 1707, and from Hatori (1976, p. 24) for 1854 and 1944.

Account in "Miho-mura yōji oboe" 『三保村用事覚』の記述

THE MIHO HEADMAN describes the 1700 tsunami as having entered "something like a very high tide" (column 2). There were seven such waves between dawn and late morning (4). Each wave rose gradually and went out "with the speed of a big river" (3). The flooded area included a bayside grove of pines (maps, p. 76, 82).

Puzzled by the waves, the headman took the precaution of sending elders and children to the high ground of Miho

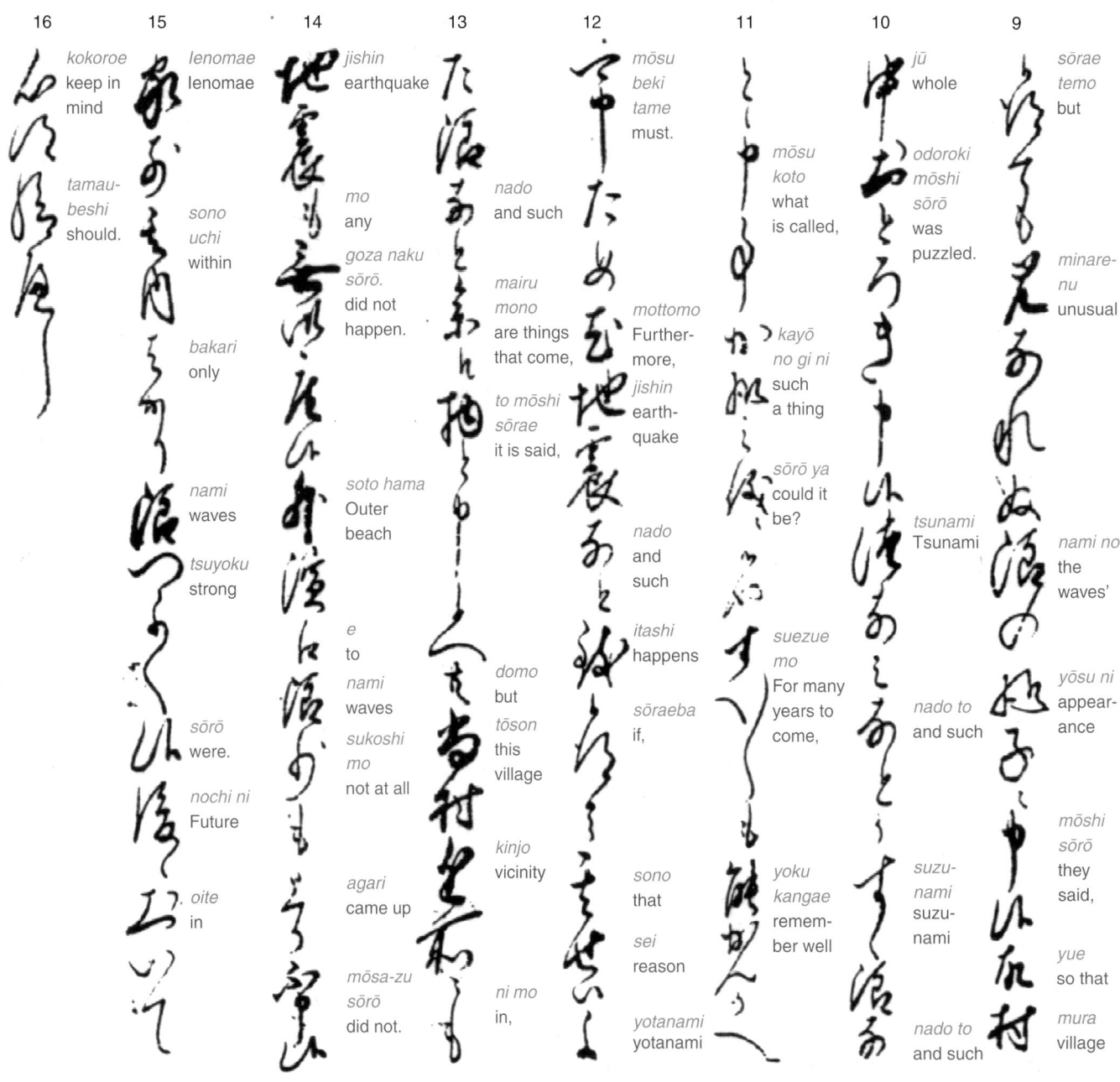

16	15	14	13	12	11	10	9
kokoroe keep in mind	*Ienomae* Ienomae	*jishin* earthquake		*mōsu beki tame* must.		*jū* whole	*sōrae temo* but
tamau-beshi should.	*sono uchi* within	*mo* any	*nado* and such	*mōsu koto* what is called,		*odoroki mōshi sōrō* was puzzled.	*minare-nu* unusual
	bakari only	*goza naku sōrō.* did not happen.	*mairu mono* are things that come,	*mottomo* Further-more,	*kayō no gi ni* such a thing		
			to mōshi sōrae it is said,	*jishin* earth-quake	*sōrō ya* could it be?		
	nami waves	*soto hama* Outer beach		*nado* and such		*tsunami* Tsunami	*nami no* the waves'
	tsuyoku strong	*e to*		*itashi* happens	*suezue mo* For many years to come,		*yōsu ni* appear-ance
	sōrō were.	*nami* waves	*domo* but	*sōraeba* if,		*nado to* and such	
	nochi ni Future	*sukoshi mo* not at all	*tōson* this village				*mōshi sōrō* they said,
	oite in	*agari* came up	*kinjo* vicinity	*sono* that	*yoku kangae* remem-ber well	*suzu-nami suzu-nami*	*yue* so that
		mōsa-zu sōrō did not.	*ni mo* in,	*sei* reason		*nado to* and such	*mura* village
				yotanami yotanami			

Formal language—*sōraeba* (8, 12), *sōrae* (9, 13), *sōrō* (9, 10, 14, 15).

Sound change at word juncture—*suezue* for *suhe suhe* (11).

Modern Japanese kana include diacritics (*dakuten*) that alter consonants, as in changing こ *ko* into ご *go*. Like some of his contemporaries, the "Oboe" writer did not use dakuten. His kana can be read, out of context, as *Eko* (instead of *Ego* in 2), *nato* (instead of *nado* in 2, 10, 12), and *shisumari* (instead of *shizmari* in 12).

The "Oboe" writer used simple phonetic script (*kana*) instead of complex ideograms (*kanji*) for *nado* (2, 10-11, 12, 13), *hodo* (4, 6), *kainaku nari* (5), *narenu* in *kikinarenu* (6) and in *minarenu* (9), *naru* (6), *shizumari* (8), *odoroki* (10), *suezue* and *kangae* (11), *tame* and *sei* (12), *yota* (12-13), *bakari, tsuyoku,* and *oite* (15). The "Oboe" script thus looks simpler than the kanji-rich columns of the other orphan-tsunami accounts (p. 34-35).

shrine (columns 6-7). The waves also perplexed elders the headman consulted (8-9). Though the headman had heard that an earthquake usually precedes a tsunami, these waves followed no earthquake felt in Miho or nearby (12-14).

Among surviving accounts of the 1700 tsunami, this one stands out as the testimony of an involved eyewitness who pondered a phenomenon beyond his life experience.

COLUMN 1

8

sōraeba but then

migi no tōri ni mentioned at right

shizumari calmed

mōshi sōrō was.

toshiyori ni To the old

kiki asked

7

o-miya shrine

e to

nigashi escape.

nokori Remain

nite wa as for,

Ienomae Ienomae

umi ni sea to

kiotsuke oru mo paying attention,

6

tsui ni kiki nare nu never heard of

nami waves

no of

agari rising

yō naru condition

hodo because of,

mura village

rōnyaku old and young

5

kainaku nari became calm.

hiru Noon

sugi after

yori from

umi sea

shizuka ni quiet

naru mo became,

shikaredomo but

4

gotoshi like.

sono That

hi no day's

yotsu hour of four

mae until,

nana- seven

do times

hodo about

agari rose,

dandan gradually

3

matsu no uchi within pine groves

made made up to

mo that far

agari reached.

sono That

hikiyō withdrawal

wa as for,

ōkawa big river

no of

hayaki speed

2

mizu water

takaku nari became high.

michishio High tide

nado no or something

yōni like

sashikomi entered,

Ego Ego

COLUMN 1

[start of entry]

jūni-gatsu Twelfth Month

kokonu-ka ninth day

ake-mutsu morning hour of six

mae before

yori from,

nami wave

Ienomae Ienomae

e to

Wada oki off Wada

yori from

11, *kangae*—Written *kangahe*.

12-13, *mottomo...mono to mōshi sōrae*—Further, it is said that if an earthquake happens, a tsunami follows.

12-13, *yotanami*—Literally, wild waves.

14, *goza naku sōrō*—Written *naku goza sōrō*.

14, *soto hama*—The beach facing the open sea.

7, *nigashi*—I advised them to escape.

7, *kiotsuke*—Romanized transliteration contains the object marker *wo*, pronounced *o.*

9. *yue*—Literally, reason.

9-10, *mura jū*—All people in the village.

10, *suzunami*—A now-unknown term that may have meant "quiet waves"

← NOTES. Columns 1-2, *Ienomae, Wada*, and *Ego*—Place names in Miho (map, p. 82).

1 and 14, *e*—Pronounced and written *e.* Signifies "to" as does *e* in 7, where written *he* へ.

3 and 7, *wa*—Topic marker written *ha* ハ.

6, *rōnyaku*—The elderly and children.

7, *o-miya*—Miho shrine (maps, p. 76, 82).

Tidal waves 津波の挙動

A tsunami that resembles a tide may sweep away buildings.

TIDE-LIKE WAVES
OF 1700 TSUNAMI

michishio
high tide

nado no
yōni
or
something
like

sashikomi
entered

p. 79, column 2

HOKUSAI'S WAVE
MISUSED AS A TSUNAMI ICON

January 18, 1996

1997

Internationalized, 2003 (p. 46)

A TOWERING, BREAKING WAVE dominates one of Hokusai's "Thirty-six views of Mount Fuji." Though an exaggerated wind wave, it has become an icon of tsunamis. Reproduced on the cover of a leading scientific journal, it represents a report on the 1700 tsunami in Japan. Adapted on a Japanese postage stamp, it commemorates diplomatic relations with a country known in Japan for its 1960 tsunami (below and opposite; see also p. 55). Simplified on American roadsides, it identifies tsunami-evacuation routes (p. 46).

The tide-like flows described by the Miho headman (quote, left) more nearly resemble a real tsunami. Tens of kilometers crest to crest, tsunami waves typically come ashore as relentlessly rising surges, like the ones captured in horrific videos from South and Southeast Asia in 2004.

The headman's words bring to mind the 1960 Chile tsunami in Japan. It entered Onagawa as waves that neither towered nor broke (below). At Ōfunato (opposite) its swift currents drove boats ashore (p. 55) while sweeping buildings off their foundations (opposite). Near Tsugaruishi and Tanabe it resembled a river in flood (photos, p. 51, 85). Such flooding probably explains the how the orphan tsunami of 1700 destroyed houses beside Miyako Bay (losses quoted, p. 48, 56).

1960 TSUNAMI AT ONAGAWA: FIRST LARGE WAVE...

Tsunami

Wave front

People fleeing

4:40 a.m.
May 24

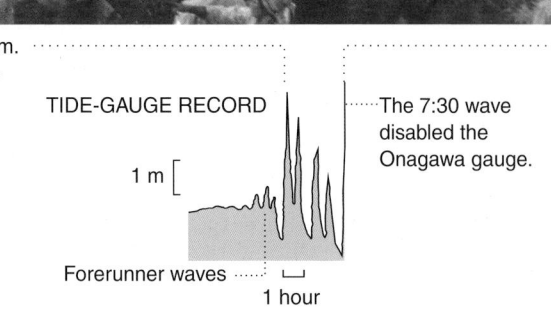

TIDE-GAUGE RECORD

1 m

The 7:30 wave disabled the Onagawa gauge.

Forerunner waves

1 hour

...AND A LATER WAVE

High-water line from 4:55 a.m.

Wave front

Debris from waves

7:30 a.m.

"THE INAPPROPRIATE ICON" is the late Doak Cox's epithet for tsunami symbols that contain or mimic Hokusai's wave. Towering breakers rarely signal a tsunami's arrival (Lander and others, 1993, p. 2; Cox, 2001). The *Nature* cover spotlights the orphan-tsunami report by Satake and others (1996). For footage of the 2004 Indian Ocean tsunami, see http://www.waveofdestruction.org/videos/.

ONAGAWA PHOTOS, attributed to M. Kondō, are from a notebook stored at the Earthquake Research Institute, University of Tokyo. An alert fireman, noting water-level changes from forerunner waves, warned residents to go to high ground before 4:40 a.m. Everyone survived (Atwater and others, 1999, p. 8).

Direction of inflow

Ripples

Person

Foundation

Courtesy of Ōfunato city. Additional view, p. 133

Selective history

THE DEVASTATION ABOVE, from the 1960 tsunami, contributed to 52 deaths in Ōfunato. The 1700 tsunami surely came ashore here as well; the two tsunamis attained similar size elsewhere in Japan (p. 48). However, the orphan tsunami of 1700 is unknown from Ōfunato, probably because of accidents of human history:

• Ōfunato belonged to a domain, Sendai-han, that kept fewer administrative records than did Morioka-han.

• Magistrates reported the 1700 tsunami incompletely even in Morioka-han. They documented its effects close to their district offices in Miyako and Ōtsuchi but neglected the damage or flooding 7 km to the south at Tsugaruishi (compare p. 38-39 with p. 52).

• The district office for Ōfunato, at Imaizumi, was 10 km away on a different bay.

• Ōfunato had less at risk in 1700. The area probably had little more than one-tenth its 1960 population. Few houses stood in 1700 on land the 1960 tsunami would rake.

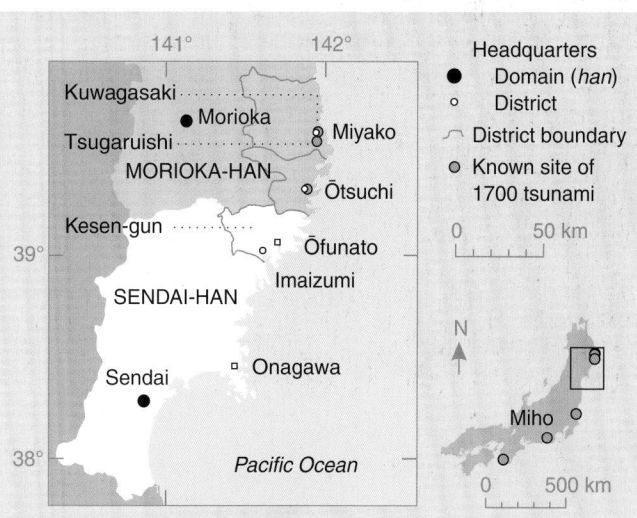

Headquarters
● Domain (*han*)
○ District
⌐ District boundary
◉ Known site of 1700 tsunami

ŌFUNATO also lacks records of the 1611 tsunami, which crested over 5 m high between Miyako and Sendai (p. 41). The village area had 1,217 residents in 1641 (Ōfunato Shishi Henshu I'inkai, 1978, appendix, p. 378), versus 11,200 in 1960. An 1822 map shows paddies and salt ponds on low ground but houses solely on uplands (Kin'no, 1981, p. 22-23).

Tsunami size 津波の高さ

At Miho, the far-traveled tsunamis of 1700 and 1960 were similarly small.

NO REPORTED DAMAGE resulted from the 1700 tsunami at Miho. To enter the pine grove, the tsunami probably rose a meter or two above ambient tide, which likely stood near mean sea level as the water repeatedly rose and fell (graphed tide curve, opposite). The 1960 tsunami similarly amounted to little at Miho, even though its largest wave coincided with a tide several tenths of a meter higher.

EVIDENCE FOR 1700
IN "MIHO-MURA YŌJI OBOE"

matsu no uchi
within pine groves
[at Ego]

TSUNAMI HEIGHT,
IN METERS AT BAY SHORE

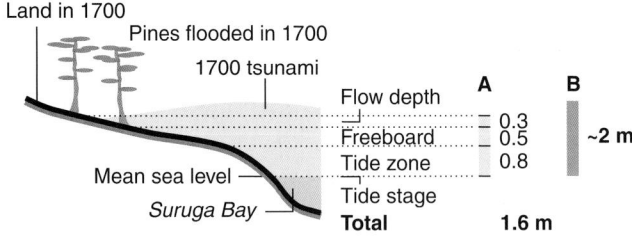

Inferred for 1700 / Measured, 1960 tsunami — A B C

MODERN LANDSCAPE

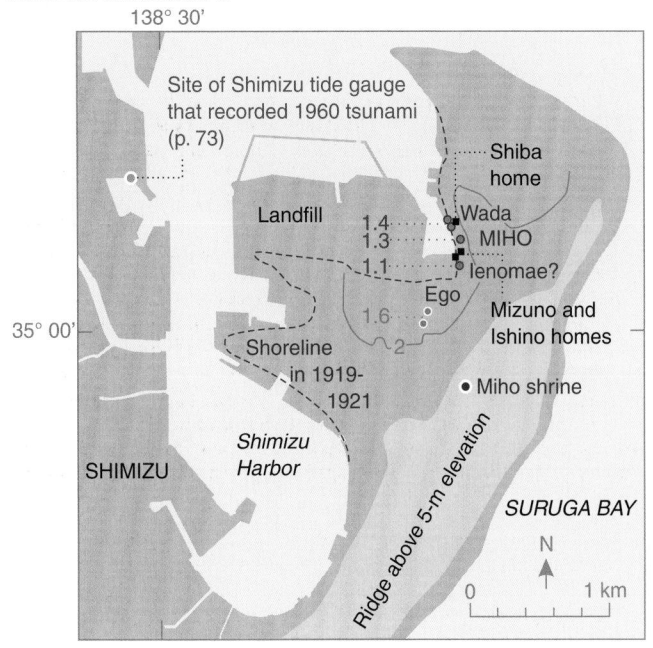

138° 30'

Site of Shimizu tide gauge that recorded 1960 tsunami (p. 73)

Shiba home

Landfill   1.4
            1.3   Wada
            1.1   MIHO
                   Ienomae?
            Ego

35° 00'   1.6
            ~2   Mizuno and Ishino homes

Shoreline in 1919-1921

• Miho shrine

SHIMIZU   *Shimizu Harbor*

SURUGA BAY

N

0   1 km

Elevation, in meters above TP on map dated 1988
~2~ 1.6● Contour and point probably above 1960 tsunami
     1.3● Point probably below level of 1960 tsunami
Ego ----- ● **Historical feature** Ienomae located approximately

EFFECTS OF 1854 EARTHQUAKE AS RECALLED IN 1893

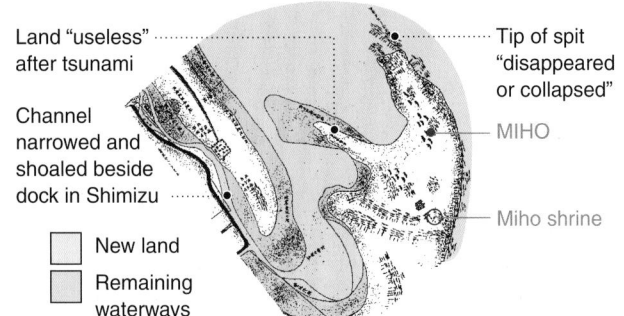

Land "useless" after tsunami

Tip of spit "disappeared or collapsed"

Channel narrowed and shoaled beside dock in Shimizu

MIHO

Miho shrine

☐ New land
☐ Remaining waterways

UPPER MAP Base traced from Kokudo Chiri'in (Geographical Survey Institute) Okitsu, Shimizu, and Shizuoka 1:25,000 quadrangles, 1996-1998. Earlier shoreline, plotted only near Miho, mapped by Rikuchi Sokuryōbu at 1:50,000, 1919-1921. Elevations from 1988 Shimizu city planning map. Tide-gauge location from Teramoto and others (1961, p. 324); Wada, Ienomae, and Ego locations from Endō and Nagasawa (1989, map 2).

1700 tsunami at Ego

A and B Height inferred with simplest assumptions

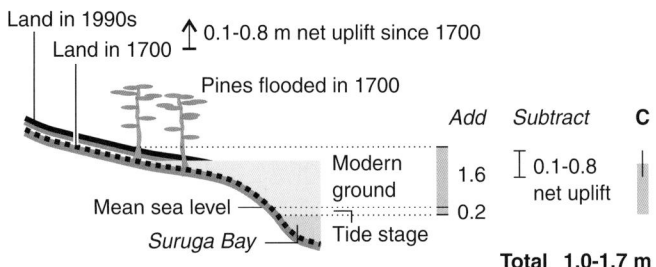

Land in 1700

Pines flooded in 1700

1700 tsunami

                                    A   B
Flow depth   0.3
Freeboard   0.5   ~2 m
Tide zone   0.8
Mean sea level
Tide stage
Suruga Bay   **Total**   1.6 m

ASSUMPTIONS
Inland change in tsunami height None (**A**). Inland descent (**B**), by analogy with the 1960 tsunami at Ōtsuchi and Shinjō (p. 65, 89).
Flow depth In pines, 0.3 m (**A**) or 0.5 m (**B**).
Freeboard Storm-wave swash excluded pines from land less than 0.5 m above highest astronomical tides.
Tide zone Highest astronomical tides 0.8 m above mean sea level.
Tide stage At 1700 mean sea level during tsunami (facing page).

C Height depends greatly on net uplift since 1700

Land in 1990s
Land in 1700   ↕ 0.1-0.8 m net uplift since 1700

Pines flooded in 1700

                        Add   *Subtract*   **C**
Modern ground   1.6   0.1-0.8
                       0.2   net uplift
Mean sea level
Tide stage
Suruga Bay

Total 1.0-1.7 m

ASSUMPTIONS
Inland change in tsunami height None.
Tide stage 0.2 m below 1700 mean sea level (facing page).
Modern ground The area of the flooded pines is now about 1.6 m above TP (map, upper left).
Net uplift 0.1-0.8 m since 1700: coseismic uplift 0.0-0.7 m in 1707 and 1.4 m in 1854; interseismic subsidence 4.8 mm/yr (250-year extrapolation of Shimizu trend, p. 65).

LOWER MAP at left originated in 1893 with officials of Shizuoka prefecture, the modern regional jurisdiction to which Miho belongs (Hatori, 1976). The validity of height **C**, from Tsuji and others (1998), depends partly on the shoreline changes reconstructed on this map. Those changes demonstrate uplift at Shimizu but not necessarily at Miho. Ishibashi (1984, p. 107) estimated that the 1854 earthquake elevated Shimizu by about 3 m. Before the earthquake, ships of 1,000-koku (180,000 liter) capacity would pull into the dock. Afterward, the channel became too narrow and shallow for such use (Hatori, 1976).

Comparisons with storms

At Miho, the 1700 and 1960 tsunamis crested below storm surges of their eras. The 1700 tsunami caused less flooding or damage in Miho than did a typhoon in 1699. Similarly, the 1960 tsunami failed to enter houses that were flooded during a storm in 1974.

In September 1699, a typhoon advanced from Kyushu to northern Honshu. In Miho it caused greater flooding and damage than did the orphan tsunami four months later. The storm surge drove villagers through waist-deep water to Miho shrine. Two persons had to be rescued from the sea; another was reported missing. On returning to their houses, villagers found high-water marks 1-2 *shaku* (0.3-0.6 m) above floor level. Three homes lost everything but their support posts. Also damaged were rice paddies, some destroyed "forever" (*eiare*). One seventy-year-old farmer, Jōzō, said he had seen his Miho paddy flooded thirteen times in 41 years. Only one or two of those floods damaged his standing crop of rice as much as did the 1699 typhoon.

Likewise for elders in modern Miho, the flooding of record accompanied a storm, not a tsunami. A storm surge on July 7, 1974 flooded the bayside homes of the Mizuno and Ishino families, in the approximate area of the neighborhood called Ienomae in 1700 (index map, opposite). Interviewed in 1999, elders of the two families recalled no other flooding of these homes, even though they lived there in 1960. The 1960 tsunami also failed to flood the home of Shiba Tsune, in nearby Wada. In her account, the water went no farther than the bayshore road in front of her house.

Adjustments for tides

When the 1700 tsunami was first noticed in Japan, the astronomical tide stood near mean sea level (left graphs below). The tsunami's midnight arrival at Kuwagasaki and Ōtsuchi coincided with a falling tide a few tenths of a meter below mean sea level. Dawn flooding at Miho and Tanabe occurred on the rising side of that low tide. Only later waves arrived with the tide above mean sea level.

In contrast, the 1960 tsunami crested during high tide (right graphs). Its largest wave nearly coincided with the peak of high tide at Miho and Tanabe. Near Miho at the Shimizu gage, the tide stood about 0.3 m above mean sea level. Because the 1960 tsunami at Miho did not exceed 1.6 m above mean sea level, it crested no more than 1.3 m above ambient tide.

TIDES The 1700 tsunami is unlikely to have piggybacked on a storm surge during its first 12 hours in Japan (p. 72). The tsunami coincided with neap tides—astronomical tides of smaller than average range (Mofjeld and others, 1997). The Pacific coast of Honshu has two daily high tides and two daily lows (Maritime Safety Agency, 1998). The mean tide range is 1-2 m. An individual tide sweeps southward, reaching Tanabe nearly 3 hours after leaving Kuwagaski.

HEIGHT OF 1960 TSUNAMI IN MIHO In the record of the Shimizu tide gauge (p. 73), the first large wave of the 1960 tsunami crested 1.3 m above TP (The Committee for Field Investigation of the Chilean Tsunami of 1960, 1961, p. 194, 371). The Mizuno and Ishino homes occupy low ground near intersections 1.1-1.3 m above TP, while the Shiba home is founded a few tenths of a meter above a road 1.4 m above TP (map, opposite). TP (Tokyo Peil) is a datum near mean sea level.

TYPHOON OF SEPTEMBER 1699 An anthology like those by earthquake historians (p. 62) identifies more than ten historical accounts of this storm (Arakawa and others, 1961). Japan's typhoon season runs from June to November and peaks in September. Storm surges account for much of the damage they cause (Arakawa and Taga, 1969, p. 129).

COMPUTED TIDE CURVES COMPARED WITH OBSERVED TIMES OF THE 1700 AND 1960 TSUNAMIS

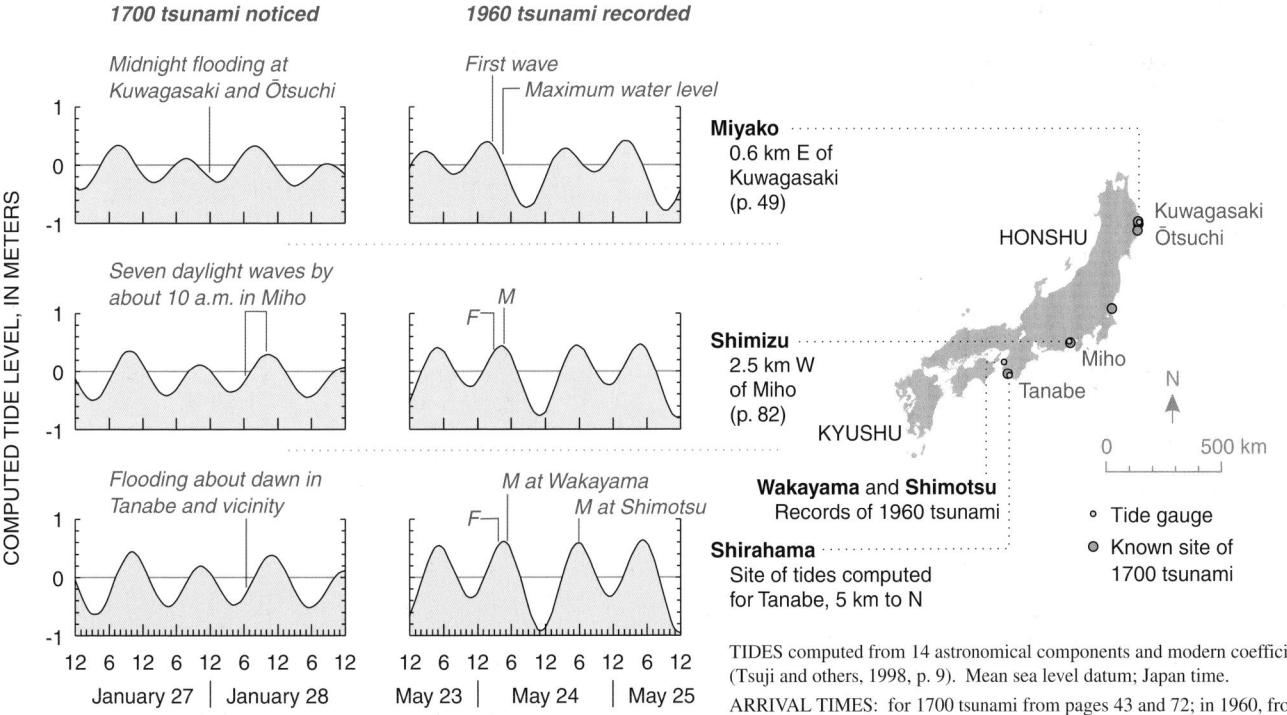

TIDES computed from 14 astronomical components and modern coefficients (Tsuji and others, 1998, p. 9). Mean sea level datum; Japan time.

ARRIVAL TIMES: for 1700 tsunami from pages 43 and 72; in 1960, from tide-gauge records of the Japan Meteorological Agency (1961, p. 34-36).

Tanabe 田辺

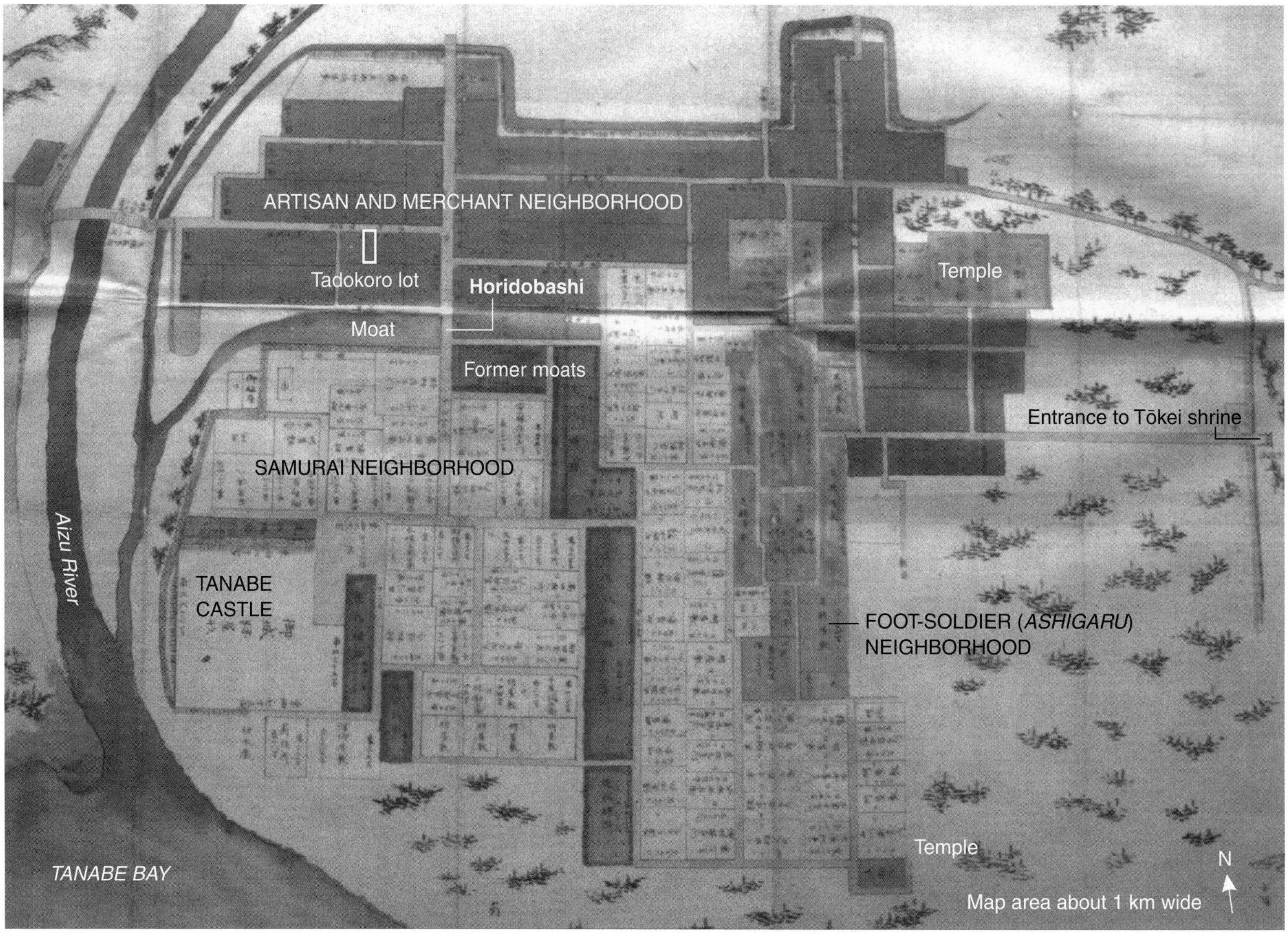

ARTISAN AND MERCHANT NEIGHBORHOOD

Tadokoro lot

Horidobashi

Moat

Former moats

Temple

Entrance to Tōkei shrine

SAMURAI NEIGHBORHOOD

Aizu River

TANABE CASTLE

FOOT-SOLDIER (*ASHIGARU*) NEIGHBORHOOD

Temple

TANABE BAY

N

Map area about 1 km wide

The castle town of Tanabe zoned large central lots for samurai and outlying neighborhoods for artisans, merchants, and foot soldiers. The town's mayor, of the merchant family Tadokoro, resided 150 meters a limit of the 1700 tsunami: the landward end of a castle moat at Horidobashi.

Volumes of "Tanabe-machi daichō" form a set of official records for the years between 1585 and 1866. The Tadokoro family kept a parallel set of private records, "Mandaiki," between 1471 and 1839. An account of the 1700 tsunami appears in both sets. Ōta Yūji, a Tanabe librarian, watched over "Daichō" in 1999 (right).

THE PICTURE MAP shows Tanabe-machi in the Hōei era (1704-1710). The map is a copy dated 1884, provided courtesy of Tanabe Municipal Library.

"TANABE-MACHI DAICHŌ," temporarily at this library when the photo at right was taken, is ordinarily held at Tōkei shrine (location on picture map above and on index map opposite). Tanabe-shi Kyōiku I'inkai (1987-1991) edited a printed version.

"MANDAIKI" can be translated as "Diary of ten thousand generations." Many of its extant volumes, including the one that covers A.D. 1700, were written in the same hand, according to librarian Ōta. In 1999 he told us that these volumes are probably copies prepared under the direction of Tanabe's seventh Tadokoro mayor (born 1758, died 1818). A printed version of "Mandaiki" runs 10,200 pages (Andō and Wakayama-ken Tanabe-shi Kyōiku I'inkai, 1991-1994).

Main points

Unusual seas off Tanabe entered a government storehouse in Shinjō, ascended a castle moat as far as Horidobashi, and flooded farmland in Atonoura, Mikonohama, and Mera (p. 86).

This inundation probably began after the 1700 tsunami's midnight arrival in Kuwagasaki (p. 43).

The tsunami probably crested 2-4 m above tide level as it crossed shores near Tanabe (p. 88-90). The flooded areas fell during subsequent earthquakes by perhaps 1 m more than they rose in between (p. 91).

Tanabe—town and district

Tanabe's mayor served also as the district mayor (*ōjōya*) of nearby villages. In this dual capacity, a mayor with the family name Tadokoro supervised the writing of "Tanabe-machi daichō" in January 1700 from his family's home in the merchant district north of Tanabe castle. There, he likely received news of the flooding in Shinjō through that village's headman, Denbe'e. Perhaps he also saw the water reach Horidobashi, 150 m from his home.

Tanabe in 1700 had 2,600 residents, probably excluding its samurai and their families. Counted among the town's commoners in 1725 were 257 fishermen, 38 fishmongers, 33 house builders, 25 innkeepers, 14 liquor merchants, 13 doctors, 3 makers of floor mats, 2 roofers, 1 stonecutter, 1 shipwright, 1 umbrella maker, and 1 merchant of palanquins (*kago*).

Shinjō in 1700 probably contained 185 houses and 240 outbuildings—structures lost to a tsunami of nearby origin in 1707 (p. 89).

Other tsunamis

As at Miho, the worst tsunamis in Tanabe originate along a plate-boundary fault off the Pacific coast of southwest Japan (p. 65, 77). An earthquake rupture 500 km long in 1707 produced a tsunami 3.5 m high in Tanabe and perhaps 8 m high in Shinjō. The fault broke again, piecemeal, in 1854 and again in the 1940s. The second of the 1854 earthquakes triggered the tsunami that led to the rice-sheaf fire in Hiro village, 40 km from Tanabe (p. 47). In Tanabe and vicinity, the 1960 tsunami from Chile crested about 3 m above ambient tide.

NOTABLE TSUNAMIS AT TANABE AND SHINJŌ SINCE 1600

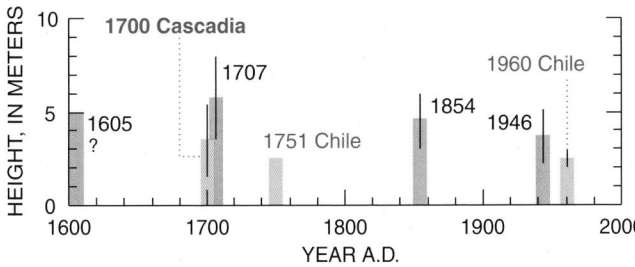

- ■ Tsunami generated nearby—Height for 1605 unknown
- ▨ Tsunami from distant source
- | Range of estimates or measurements

PLATE-TECTONIC SETTING

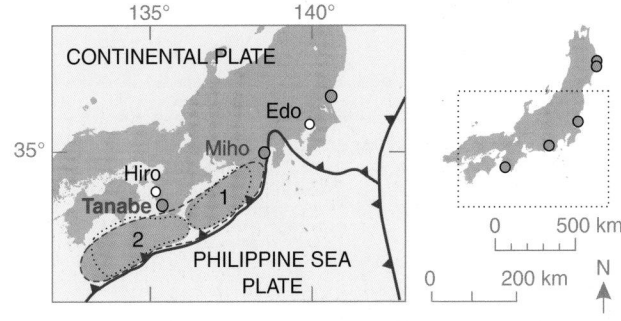

➤ Upper edge of plate-boundary fault (p. 8, 65, 77)

Rupture area of great earthquake on plate-boundary fault
- 1707—Areas 1 and 2 combined
- 1854—32 hours apart, area 1 first
- 1944 (1) and 1946 (2)—M 8.1, 8.1; smaller than 1854
- ● Known site of 1700 Cascadia tsunami

TANABE AND VICINITY

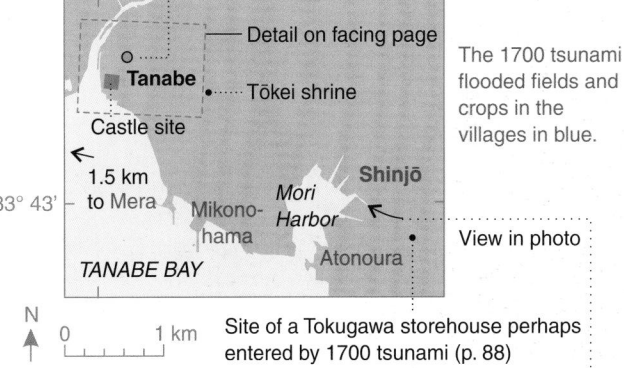

The 1700 tsunami flooded fields and crops in the villages in blue.

1960 CHILE TSUNAMI IN SHINJŌ

Inflow from Mori Harbor

EARTHQUAKES AND TSUNAMIS Ando (1975) and Ishibashi (1981) estimated rupture areas of the 1707, 1854, 1944, and 1946 earthquakes. The Tanabe map is traced from Kokudo Chiriin (Geographical Survey Institute), Kii Shirahama and Kii Tanabe 1:25,000, 1990 and 1996. Shinjō Kōminkan, a community center, provided the above photo (location, p. 89). The tsunami heights are from Watanabe (1998, p. 71, 80, 96, and 136), Japan Meteorological Agency (1961, p. 192), Yoshinobu (1961), and our interpretation of a "Mandaiki" account of slight flooding in Shinjō during the 1751 tsunami (footnote, p. 54).

TOWN AND DISTRICT Takeuchi (1985b, p. 658) lists commoners' occupations and gives Tanabe's population as 2,516 in Kanbun 7 (1667) and 2,720 in Kyōho 10 (1725). The totals exclude children under nine. Kishi Akinori, a local historian, told us in 1999 about the Tadokoro mayors and the Shinjō headman.

Account in "Tanabe-machi daichō"　『田辺町大帳』の記述

SEAS ROSE STRANGELY near Tanabe around dawn of the 8th day (columns 1-2). The water entered a Tokugawa storehouse in Shinjō village and other buildings, too. In addition, the water damaged crops and fields in the Atonoura area of Shinjō. Within Tanabe itself, not far from the writer, the water ascended a castle moat as far as Horidobashi (3-4).

This account comes from the water-stained volume pictured at right.

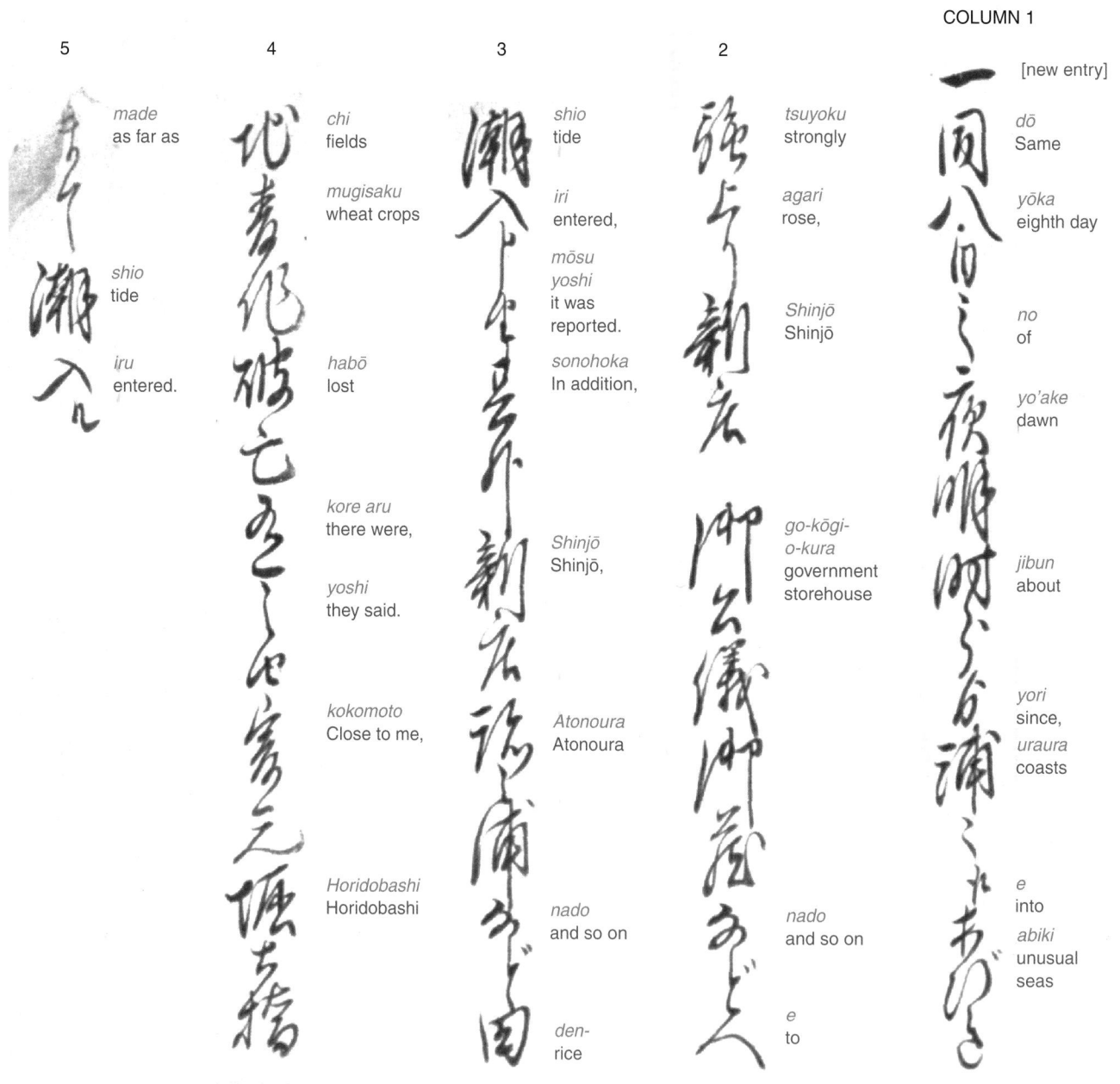

COLUMN 1

5	4	3	2	1 [new entry]
made as far as	*chi* fields	*shio* tide	*tsuyoku* strongly	*dō* Same
shio tide	*mugisaku* wheat crops	*iri* entered,	*agari* rose,	*yōka* eighth day
iru entered.	*habō* lost	*mōsu yoshi* it was reported.	*Shinjō* Shinjō	*no* of
		sonohoka In addition,		*yo'ake* dawn
	kore aru there were,		*go-kōgi-o-kura* government storehouse	*jibun* about
	yoshi they said.	*Shinjō* Shinjō,		*yori* since,
	kokomoto Close to me,	*Atonoura* Atonoura		*uraura* coasts
	Horidobashi Horidobashi	*nado* and so on	*nado* and so on	*e* into
		den- rice	*e* to	*abiki* unusual seas

3, *shio*—Composite symbol for "water" and "morning" (p. 40).

3, *Shinjō Atonoura*—An account from "Mandaiki," from the year 1707, treats Atonoura as part of Shinjō village (Tokyo Daigaku Jishin Kenkyūsho, 1981, p. 326). Its passage on the 1700 tsunami mentions damage to fields not only in Atonoura but also in Hama (that is, Mikonohama) and Mera. See page 85 for an index map and page 84 for notes on "Mandaiki."

possibilities include tide, storm, wind, and changes in atmospheric pressure, in addition to tsunami (p. 40; Hibiya and Kajiura, 1982; Yanuma and Tsuji, 1998).

2, *go-kōgi*—*go*, honorific; *kōgi*, public affairs (Berry, 1982, p. 158; Hall, 1991, p. 19). Refers here to the branch of the Tokugawa clan that ruled Wakayama-han, of which Tanabe was a part.

2, *e*—Pronounced *e*, written *he*, signifies "to."

← NOTES. Column 1, *dō*—Same year and month as in preceding entry.

1, *uraura*—Repeat symbol ⟨ makes *ura* plural. The unusual seas occurred along more than one part of the coast near Tanabe.

1, *e*—Written *e*, means "into."

1, *abiki*—The term refers to unusual seas without necessarily implying their cause. From usage that varies with region, and perhaps also with time,

VOLUME OPENED TO TSUNAMI ENTRY

TEXT OF INSIDE TITLE PAGE

2   Column 1

dō same [era]		*Genroku* Genroku	
jūni twelve			
		hachi eight	
tsuchinoto younger brother of earth		*kinoto* younger brother of wood	
u rabbit		*i* boar	
made to,		*nen* year	
shirusu recorded		*yori* from,	
kore this.			

Genroku 8 began in 1696; Genroku 12 ended in 1700 (p. 42).

TSUNAMI ENTRY

Water damage

Durable history

ACCOUNTS OF THE 1700 TSUNAMI were brushed onto *washi*. Strengthened by fibers of bark, washi has served as writing paper, screens, windows, lantern covers, and even clothing. Its use in Japan predates the 1700 tsunami by more than 1,000 years.

The tsunami accounts have survived water and bookworms. Water erased an edge of the account in "Tanabe-machi daichō" (above). Worms known as *shimi* (紙魚 paper fish) explored most of the source documents, including Morioka-han "Zassho" (below). Additional bookworms leaf through sturdy pages at right.

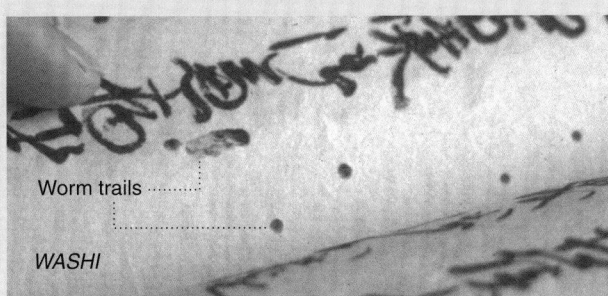

Worm trails

WASHI

Bookworm passages perforate the volume of Morioka-han "Zassho" for the year Genroku 12 (above; entire volume, p. 42). Such trails also cross the map folds on page 32 and riddle the book cover on page 66. At upper right, earthquake historians devour durable documents in Tanabe.

TANABE LIBRARIAN Ōta Yūji identified as *kōyawashi* the paper used by the Tadokoro mayor who copied "Mandaiki" in the late 1700's or early 1800's. This paper takes its name from manufacture in Kōya, 60 km north of Tanabe.

IN MORIOKA, librarian Konishi Hiroaki surmised that the washi in "Morioka-han zassho" was imported from the south, for want of suitable fiber in northern Honshu.

ON WASHI'S MANUFACTURE and use, see All Japan Handmade Washi Association (1991). Boudonnat and Kushizaki (2003, p. 187) discuss the paper's antiquity. Chamberlain (1905, p. 360), introducing Westerners to "Things Japanese," reported that washi "lends itself admirably to the native brush, but not to our pointed pens, which stick and splutter in its porous fibre." Paper manufacturing in North America's English colonies probably began in 1700, in Philadelphia (Trager, 1992, p. 271).

Tsunami size near a storehouse 御蔵付近の津波の高さ

The 1700 tsunami probably reached heights of several meters in Shinjō.

CROSSING THE SHORE on its way to the government storehouse in Shinjō, the 1700 tsunami crested at least 2 m above tide (estimate **A**). A height of 4 m is reasonable if the storehouse stood on low ground at least half a kilometer inland, and if the tsunami height descended inland as it did in 1960 (**B**; heights in 1960 mapped on facing page).

The tsunami rose more than 5 m if the storehouse stood on high ground identified in Shinjō oral tradition (**C**). That tradition places a bygone government storehouse at the site in the photo below (map, opposite). However, this site was not necessarily the one flooded in 1700: Shinjō had more than one government storehouse in 1707, when a tsunami destroyed two of them (box, facing page).

STOREHOUSE SITE IN SHINJŌ

Road overtopped by 1946 tsunami but not by 1960 tsunami

STOREHOUSE FLOODING IN 1700

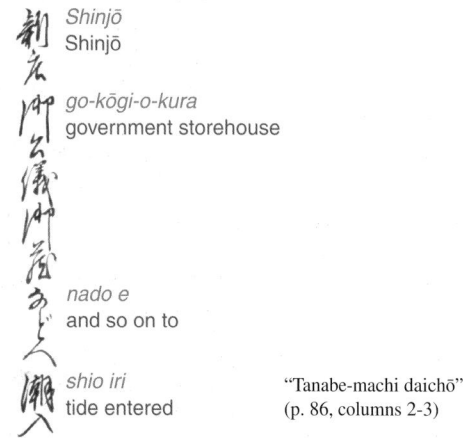

Shinjō
Shinjō

go-kōgi-o-kura
government storehouse

nado e
and so on to

shio iri
tide entered

"Tanabe-machi daichō"
(p. 86, columns 2-3)

AT THE STOREHOUSE SITE in the photo above, the family of Matsuzaki Tomiji built a house early in the 20th century. Mr. Matsuzaki, born in 1926, told us in 1999 that he saw the 1960 tsunami stop short of this house and also the street fronting it. Mr. Matsuzaki also recalled being told that this street was crossed by the tsunami from the region's great 1946 earthquake (p. 85).

ESTIMATE **C** is from Tsuji and others (1998).

HEIGHT DATUMS. Tide tables of the Maritime Safety Agency (1998) list the highest astronomical tide at Tanabe as 1.04 m above mean sea level. TP is a datum near mean sea level.

SUMMARY OF TSUNAMI HEIGHTS

At edge of Mori Harbor, in meters above ambient tide. Inferred for 1700 (diagrams below), measured for 1960 (map opposite).

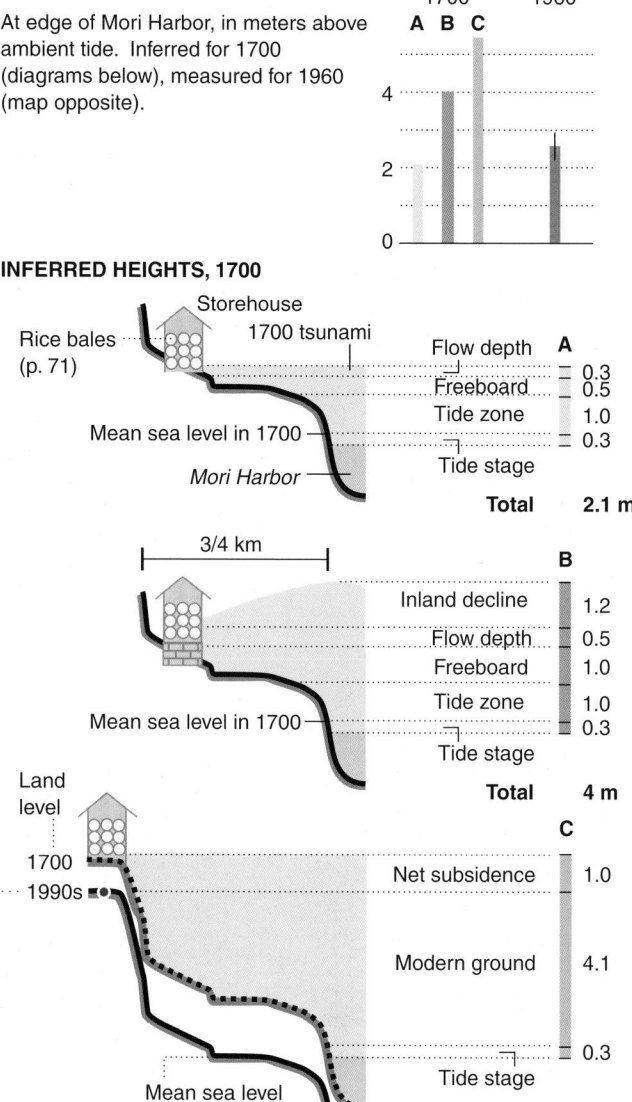

INFERRED HEIGHTS, 1700

A

Flow depth	0.3
Freeboard	0.5
Tide zone	1.0
Tide stage	0.3
Total	**2.1 m**

3/4 km

B

Inland decline	1.2
Flow depth	0.5
Freeboard	1.0
Tide zone	1.0
Tide stage	0.3
Total	**4 m**

C

Net subsidence	1.0
Modern ground	4.1
Tide stage	0.3
Total	**5.4 m**

ASSUMPTIONS

Flow depth 0.3 m for storehouse with minimal foundation (**A**), 0.5 m for foundation typical of traditional storehouses in former samurai neighborhoods of Tanabe (**B**; storehouse photo, p. 108).

Freeboard To keep government rice above waves during storm surges, storehouse was sited 0.5 m (**A**) or 1.0 m (**B**) above highest astronomical tides.

Tide zone Highest astronomical tides were 1.0 m above mean sea level, the modern value for Tanabe listed in tide tables. Relevant to **A** and **B** only.

Tide stage When storehouse flooded at or before dawn, tide stood 0.3 m below 1700 mean sea level (p. 83). Used in all estimates.

Inland decline Tsunami crest descended inland in 1700 as much as it did in 1960 (map on facing page). **B** only.

Net subsidence 1.0 m since 1700 (p. 91). **C** only.

Modern ground The storehouse site now stands 4.1 m above TP (photo at left; map and airphoto on facing page). **C** only.

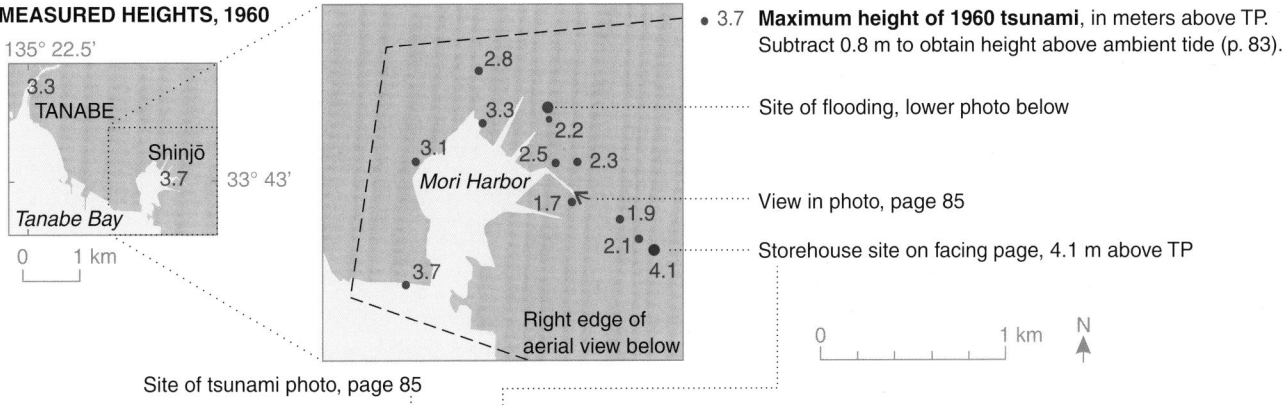

MEASURED HEIGHTS, 1960

135° 22.5'

3.3
TANABE

Shinjō
3.7

Tanabe Bay

33° 43'

0   1 km

2.8

3.3

2.2

3.1   2.5   2.3

Mori Harbor

1.7   1.9

2.1

3.7   4.1

Right edge of
aerial view below

Site of tsunami photo, page 85

• 3.7   **Maximum height of 1960 tsunami**, in meters above TP.
Subtract 0.8 m to obtain height above ambient tide (p. 83).

Site of flooding, lower photo below

View in photo, page 85

Storehouse site on facing page, 4.1 m above TP

0   1 km   N

Breakwater built in 1965

Mori Harbor

TANABE BAY

Modern Shinjō sprawls across lowlands beside
Mori Harbor, in an aerial view from the 1990s.
Storehouse site overlooks a field that the 1960
tsunami partly flooded.

Hoof-deep water of the 1960 tsunami covers a street 250 meters
from the harbor.

1960 TSUNAMI HEIGHTS. The heights plotted above, at left, are from a
regional report by the Japan Meteorological Agency (1961, p. 192); at right, from
a local survey of Shinjō headed by a schoolteacher, Yoshinobu Eiji. Mr. Yoshinobu
sought comparisons with the 1946 tsunami, whose heights he had previously
surveyed with a hand level. After the 1960 tsunami, Tanabe's mayor provided him
with the services of Fujino Fumitada, a licensed surveyor, and Otani Yasuzō, an
assistant to Mr. Fujino. Mr. Yoshinobu pointed them to the points he had measured
in 1946, as well as to levels reached by the Chilean tsunami. The men surveyed
for three days. Their findings appear on pages 20 and 23 of Yoshinobu's report
(which also describes the building of a breakwater at the entrance to Mori Harbor
in 1965). Okamoto Yoshihiko of Tanabe's city office provided us with a copy of
the report and with the airphoto above. The lower photo comes from a collection
at Shinjō Kōminkan (a community center), courtesy of Kashiwagi Tamio.

1960 TSUNAMI DAMAGE. Except for the south shore of Miyako Bay (p. 51),
no recorded site of the 1700 tsunami in Japan suffered more damage from the
1960 tsunami than did the area around Mori Harbor. The area's losses, compiled
by Wakayama Prefecture and reported by the Japan Meteorological Agency (1961,
p. 193), totaled 1.66 million yen (U.S. $2,700 in 1960, or $16,900 adjusted for
inflation to 2003; http://www.bls.gov/cpi/).

Confounding clue from 1707

BOTH TADOKORO ACCOUNTS of the 1700
tsunami mention the flooding of one *o-kura*, or
government storehouse, in Shinjō. Oral tradition in
Shinjō places a government storehouse on high ground
in a neighborhood called *o-kura yashiki* (government-
storehouse district; photo on facing page). But the
storehouse flooded in 1700 and that remembered by
tradition are not necessarily the same, as shown by an
account of the 1707 tsunami in Shinjō.

The great Hōei era earthquake and tsunami of
October 28, 1707 devastated Shinjō. Losses there,
reported in Tadokoro documents, included 185 houses,
196 sheds, 40 cattle shelters, and five private
storehouses (*kura*). The losses also included two *o-
kura*—two government storehouses. Which, if either,
did the 1700 tsunami enter? Which corresponds to the
storehouse in Shinjō's oral tradition?

ADDITIONAL LOSSES IN 1707   The great Hōei-era earthquake and
tsunami of October 28, 1707 devastated Tanabe as well. In that castle
town, 24 persons died, 138 houses and 75 storehouses collapsed, 154
houses and 6 storehouses were washed away, and 119 houses suffered
severe damage (*taiha*). Among the houses destroyed was that of the
Tadokoro mayor (location, p. 84, 90). Left standing, but entered by the
water, was the family's adjacent storehouse that likely held "Tanabe-
machi daichō" (p. 86) and "Mandaiki." Salt water soaked the Tadokoro
records; these dried the following week.

SOURCE DOCUMENTS   "Shinshu Nihon jishin shiryō" (p. 62) cites
Tadokoro documents as the authority on the 1707 damage in Shinjō,
while quoting "Mandaiki" on the losses in Tanabe (Tokyo Daigaku Jishin
Kenkyūsho, 1981, p. 117, 135).

Tsunami size near Tanabe Bay 田辺湾付近の津波の高さ

Of modest height, the 1700 tsunami flooded bayside fields but not the mayor's house.

SIMPLE ASSUMPTIONS about reported damage to fields and crops yield bayshore heights up to 3 m for the 1700 tsunami near Tanabe (**A** and **B**). In Tanabe proper, near the home of the Tadokoro mayor, the water ascended a castle moat without reportedly overtopping its rim. Perhaps, as assumed in **C**, the tsunami approached the moat rim in the area of the motorcycle at lower right.

TSUNAMI HEIGHT, IN METERS ABOVE AMBIENT TIDE

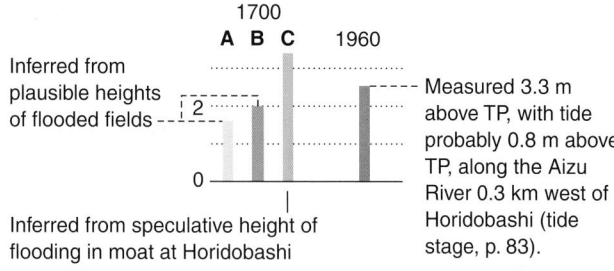

Inferred from plausible heights of flooded fields

1700
A B C 1960

Measured 3.3 m above TP, with tide probably 0.8 m above TP, along the Aizu River 0.3 km west of Horidobashi (tide stage, p. 83).

Inferred from speculative height of flooding in moat at Horidobashi

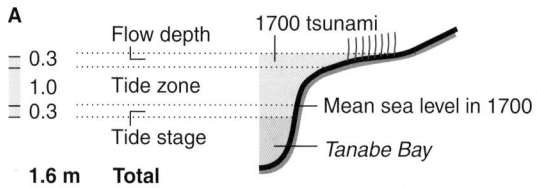

A
0.3 — Flow depth
1.0 — Tide zone
0.3 — Tide stage
1700 tsunami
Mean sea level in 1700
Tanabe Bay
1.6 m Total

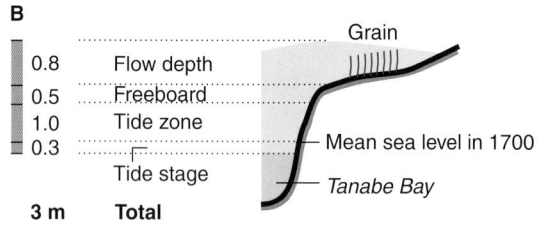

B
0.8 — Flow depth
0.5 — Freeboard
1.0 — Tide zone
0.3 — Tide stage
Grain
Mean sea level in 1700
Tanabe Bay
3 m Total

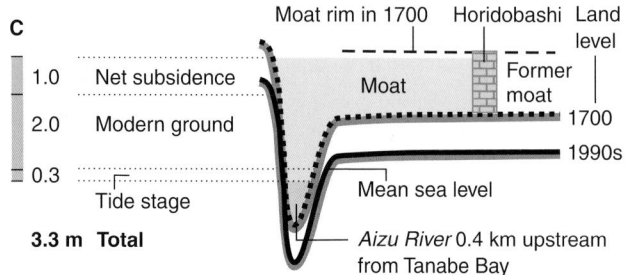

C
1.0 — Net subsidence
2.0 — Modern ground
0.3 — Tide stage
Moat rim in 1700 Horidobashi Land level
Former moat
Moat
1700
1990s
Mean sea level
Tide stage
Aizu River 0.4 km upstream from Tanabe Bay
3.3 m Total

ASSUMPTIONS
Flow depth At least 0.3 m to damage grain (**A**); higher at shore (**B**).
Freeboard Kept fields above waves during most storm surges (**B**).
Tide zone Highest astronomical tides 1.0 m above mean sea level, as listed for Tanabe in modern tide tables (**A** and **B**; datum, p. 88).
Tide stage When fields and moat flooded, tide stood 0.3 m below 1700 mean sea level (p. 83). Used in all estimates.
Net subsidence 1.0 m since 1700 (facing page). Used in **C** only.
Modern ground Tsunami approached level of moat rim, now an intersection about 2 m above TP (**C**; photo, right).

Estimate **C** from Tsuji and others (1998). TP, a datum near mean sea level.

FLOODING DESCRIBED IN TADOKORO DOCUMENTS

FIELDS IN VILLAGES*

denchi
rice paddies

mugisaku
wheat crops

habō
lost

* Mera, Mikonohama, and Atonoura (p. 86, footnote on column 3)

MOAT IN TANABE

kokomoto
Close to me,

Horidobashi
Horidobashi

made
as far as

shio iru
morning tide entered.

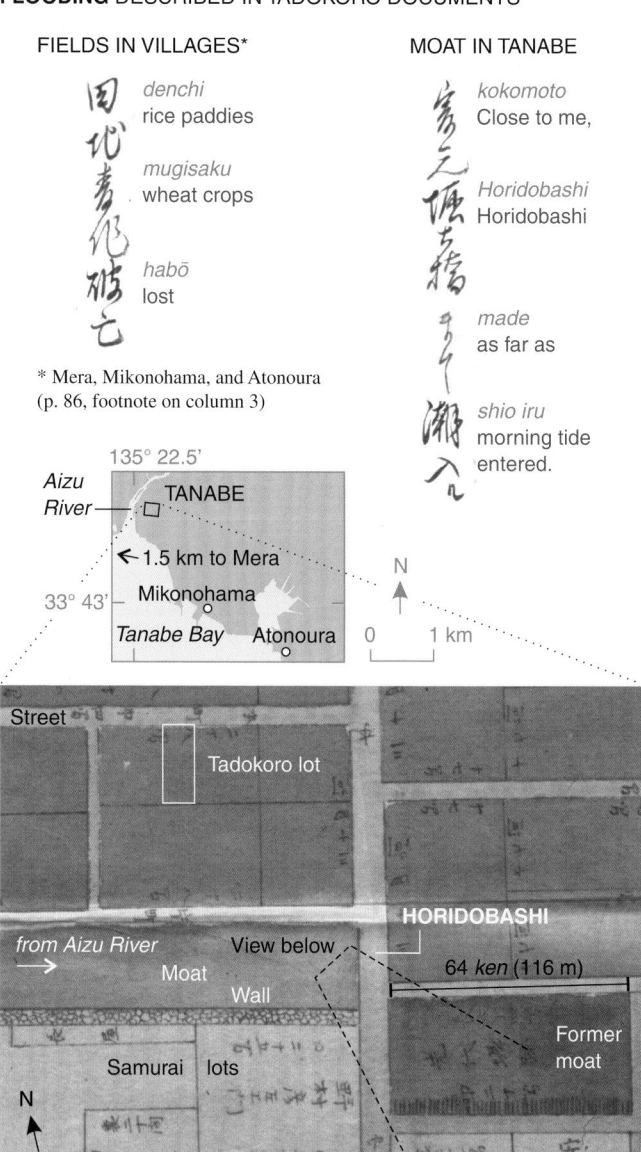

135° 22.5'
Aizu River
TANABE
← 1.5 km to Mera
33° 43'
Mikonohama
Tanabe Bay
Atonoura
N
0 1 km

Street
Tadokoro lot
from Aizu River →
Moat
View below
Wall
Samurai lots
N
HORIDOBASHI
64 *ken* (116 m)
Former moat

Site of:
Former moat   HORIDOBASHI   Moat

On the picture map from 1704-1711 (p. 84), the former moat is labeled *sendai hori ato* (moat in preceding generation). The neighborhood in the photo is known today as Horidobashi. Officials of the General Affairs Section, City of Tanabe, showed us the location of the Tadokoro lot.

Sawtooth cycles 地震サイクル

Tanabe sinks during great earthquakes and probably rises between them.

GREAT SUBDUCTION EARTHQUAKES lower land at Tanabe. The subsidence probably punctuates cycles that plot like sawteeth (below). The cycles result from stick-slip subduction (p. 10), as do the land-level changes at Chile, Alaska, and Cascadia (p. 11, 14-25). The cycles' net effect at Tanabe adds 1 m of inferred tsunami height in estimate **C** (opposite). At Tanabe since 1700, subsidence during earthquakes probably exceeded the area's gradual uplift.

COASTAL LAND-LEVEL CHANGE DURING EARTHQUAKES ALONG THE NANKAI TROUGH

1707 (largest)   1854 (larger than 1946)   1946 (M 8.1)

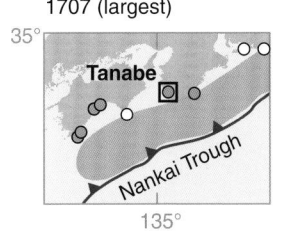 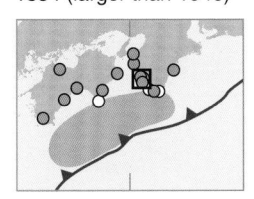 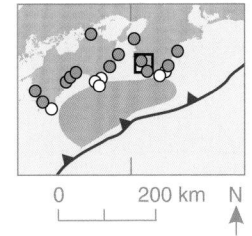

◉ **Subsidence**   Maximum of 2 m in 1707, 1.1 m on 24 December 1854, and 1.2 m in 1946

○ **Uplift**   Maximum of 2.0-2.5 m in 1707, 1.5 m in 1854, and 1 m in 1946

Earthquake rupture area   Mainly on plate-boundary fault of Nankai subduction zone

▶ **Plate-boundary fault**, seaward edge

INFERRED CYCLES OF LAND-LEVEL CHANGE AT TANABE

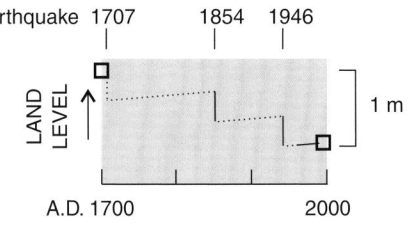

Gradual uplift between earthquakes, at 0.8 mm/yr
— Estimated from tide-gauge records 1967-1995 (p. 65)
····· Extrapolated to earlier intervals

Sudden subsidence during earthquakes, 0.4 m per event
| Measured in 1946, estimated for 1854
⋮ Reported in 1707 but amount unknown

SUBSIDENCE ESTIMATES for the earthquakes in 1707, 1854, and 1946 are from Usami (1996, p. 303). Tanabe subsided about 0.4 m during the 1946 earthquake according to a comparison of geodetic benchmarks leveled before and after the event (Thatcher, 1984, p. 3090).

THE UPLIFT RATE of 0.8 mm/yr is the average estimated by Ozawa and others (1997, p. 13) from tide-gauge records at Shirahama, 5 km southwest of Tanabe (graph, p. 65). Ando (1975) and Savage and Thatcher (1992) report additional evidence for historical land-level change at the Nankai Trough.

Firsts

INSTRUMENTAL RECORDS of confirmed tsunamis begin with a pair of Japanese wave trains that registered on tide gauges in Oregon and California on December 23 and 25, 1854.

The December 23 tsunami, originating off Miho (p. 77), in turn yielded pioneering estimates of Pacific Ocean depths. Its wave train was noticed on a San Diego marigram by the gauges' installer, William P. Trowbridge. Suspecting a submarine earthquake, Trowbridge notified Alexander Dallas Bache, head of the U.S. Coast Survey. Months later, Bache learned that an earthquake and tsunami had struck southwest Japan about 9 a.m. local time on December 23, 1854. Bache combined this news with the marigrams and with wave physics to estimate the average ocean depth between Shimoda (location, p. 77) and San Francisco. His estimate, 4.1-4.6 km, scarcely differs from today's, 4.7 km.

The second wave train originated on December 24, Japan time, off Tanabe (map, above). Its effects in Hiro village, 35 km northwest of Tanabe, inspired the story that brought "tsunami" into the English language (p. 47).

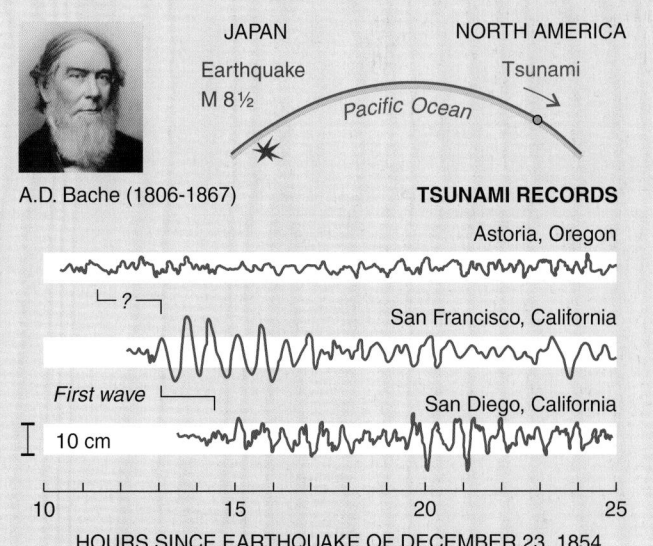

A.D. Bache (1806-1867)

THE GAUGED WAVES, as presented by Bache (1856), are plotted relative to ambient tide (like the similated waves on p. 37). Theberge (2003) tells of the gauges and Bache's estimates; Lander and others (1993, p. 40), early tsunami recordings. The photo accompanies a eulogy at http://www.history.noaa.gov/giants/bache.html. Astoria plotted on map, p. 125.

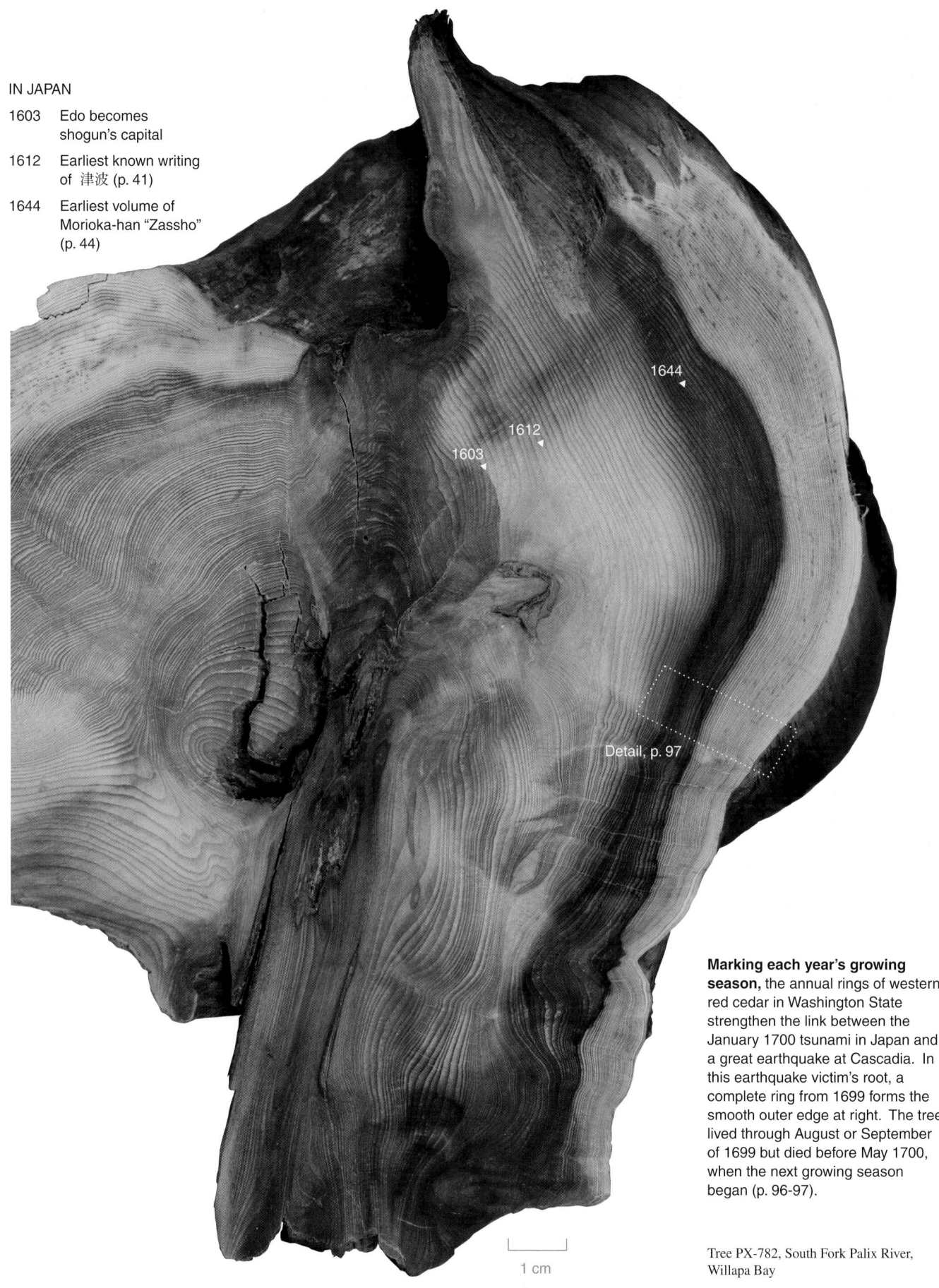

IN JAPAN

1603 Edo becomes
      shogun's capital

1612 Earliest known writing
      of 津波 (p. 41)

1644 Earliest volume of
      Morioka-han "Zassho"
      (p. 44)

1644 ◂

1612 ◂

1603 ◂

Detail, p. 97

Marking each year's growing season, the annual rings of western red cedar in Washington State strengthen the link between the January 1700 tsunami in Japan and a great earthquake at Cascadia. In this earthquake victim's root, a complete ring from 1699 forms the smooth outer edge at right. The tree lived through August or September of 1699 but died before May 1700, when the next growing season began (p. 96-97).

Tree PX-782, South Fork Palix River, Willapa Bay

1 cm

Part 3
The orphan's parent 津波の親地震

A TRANS-PACIFIC REUNION took place in 1996. Orphaned for nearly 300 years, the 1700 tsunami in Japan was reunited, on the pages of a scientific journal, with an earthquake and tsunami in North America (p. 94-95). The orphan dated the earthquake to the evening of January 26, 1700 (p. 42-43) and gave its approximate size as magnitude 9.

Today the 1700 tsunami is securely linked to a giant North American earthquake. The tie was strengthened in 1997 by tree-ring dating that narrowed the time window for a great Cascadia earthquake to the months between August 1699 and May 1700 (opposite; p. 96-97). The earthquake's enormity was confirmed in 2003 through improved estimates of the orphan tsunami's size and from computer simulations of Cascadia earthquakes and of the tsunami itself (p. 98-99). The tsunami's written record in Japan has become clearer, too, with discovery in 1998 of the Miho headman's account, authentification in 2002 of the Nakaminato shipwreck certificate, and explanation in 2004 of a discordant date from Tsugaruishi (p. 62).

The fault that broke in 1700 has been reloading for future Cascadia earthquakes. If the fault behaves as it has the last few thousand years, the earthquakes will happen sporadically at intervals ranging from a few centuries to a millenium (p. 100-101). Sometimes the fault may break along its entire length; at other times it may break piecemeal.

Today, public officials are taking steps to prepare coastal communities for Cascadia tsunamis, and engineers are using new seismic-hazard maps that allow for shaking from Cascadia earthquakes as large as magnitude 9 (p. 102-105). The story of the orphan tsunami of 1700 continues through these public-safety efforts.

By elimination 消去法によって

No other place rivals Cascadia as the orphan tsunami's source.

POTENTIAL SOURCES OF THE 1700 TSUNAMI

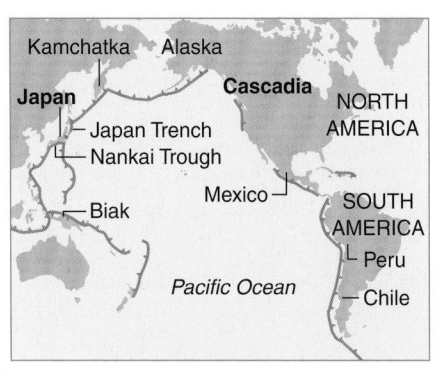

Subduction zone Line shows upper edge of plate-boundary fault. Teeth point down fault (p. 8).

MIHO'S HEADMAN WONDERED what made the 1700 tsunami (p. 78, columns 9-16). That mystery grew as 20th-century historians collected accounts of its orphan waves from Kuwagasaki to Tanabe (p. 54, 62). Geologic clues in North America, summarized in Part 1, show that the tsunami could have originated at the Cascadia subduction zone. But might the waves' real source lie elsewhere?

There is no reason to believe that the 1700 tsunami began in the seas directly off Japan. No precursory earthquake was felt along the Japan Trench at Tsugaruishi or along the Nankai Trough at Miho (p. 54). Nor did the tsunami coincide with a Japanese storm (p. 72).

Other potential sources around the Pacific Rim conflict with the tsunami's year or height. South American catalogs give sources for tsunamis recorded in Japan in 1687, 1730, and 1751, but not for any tsunami in 1700 (p. 54). The 20th century's third-largest earthquake, in Kamchatka, produced a tsunami in Japan with heights of a few meters in the north but less than 1 m in the south (graph, right; map, opposite). The 1964 Alaska tsunami, from the century's second-largest earthquake, radiated mainly off the long side of the area of a sea-floor uplift—southeastward, away from Japan—and therefore crested no more than 1 m high in Japan. An eastern Indonesian tsunami in 1996 amounted to little in Japan except on tips of southern peninsulas.

A CASCADIA SOURCE for Japan's orphan tsunami of 1700 was proposed by Satake and others (1996). Kerr (1995) and Kanamori and Heaton (1996) commented on the breakthrough.

SPANISH AMERICA in 1700 included the Pacific coast from Peru to central Chile (Haring, 1963)—sources of the tsunamis recorded in Japan in 1586, 1687, 1730, and 1751 (p. 54). Spaniards described 19 tsunami-causing earthquakes in Peru and Chile between 1650 and 1750 (Lomnitz, 1970; Lockridge, 1985). Among these, the event closest to 1700 was one that damaged northern Chile in 1705. In Mexico, shaking on June 30, 1700 was recorded both on the Pacific coast and inland, and other temblors were recorded inland on September 29, 1699 and on March 30, 1700 (García and Suárez, 1996, p. 106).

TSUNAMI HEIGHTS IN JAPAN

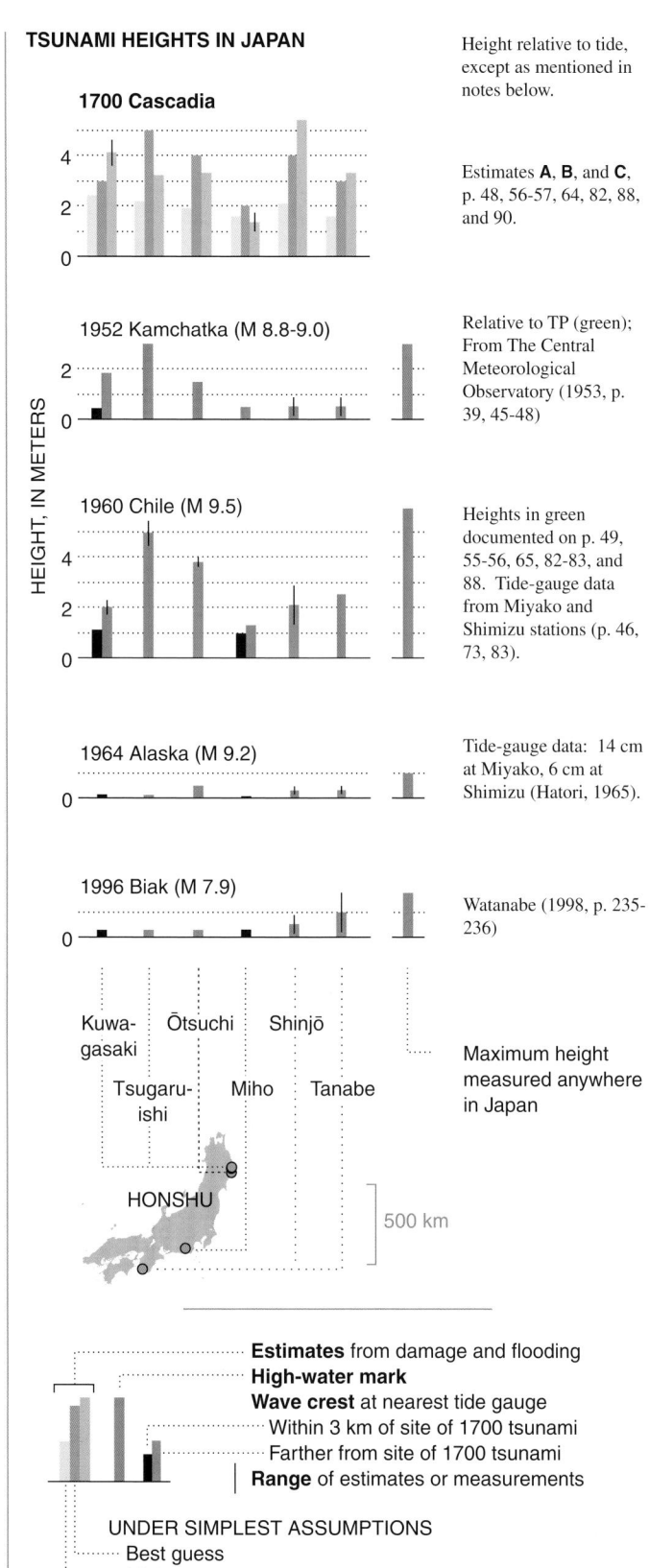

Height relative to tide, except as mentioned in notes below.

Estimates **A**, **B**, and **C**, p. 48, 56-57, 64, 82, 88, and 90.

Relative to TP (green); From The Central Meteorological Observatory (1953, p. 39, 45-48)

Heights in green documented on p. 49, 55-56, 65, 82-83, and 88. Tide-gauge data from Miyako and Shimizu stations (p. 46, 73, 83).

Tide-gauge data: 14 cm at Miyako, 6 cm at Shimizu (Hatori, 1965).

Watanabe (1998, p. 235-236)

1952 KAMCHATKA TSUNAMI: HIGH WATER MAINLY NORTH

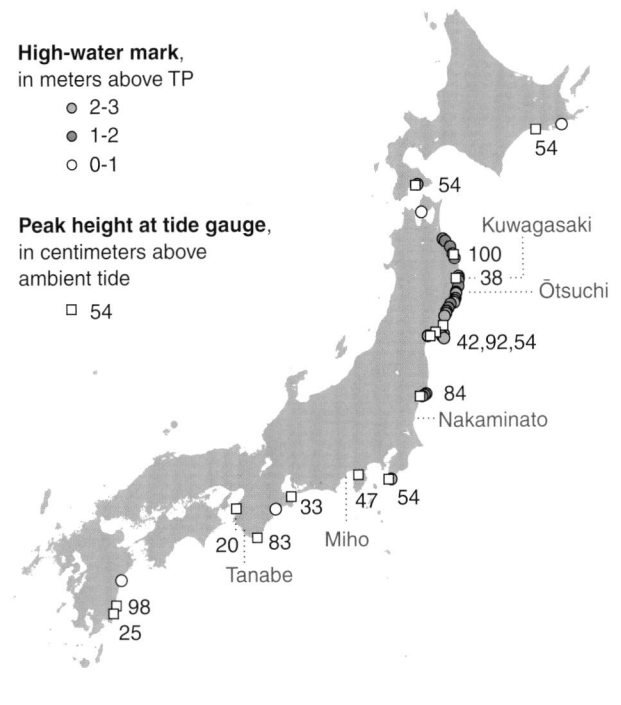

High-water mark,
in meters above TP
- ◉ 2-3
- ● 1-2
- ○ 0-1

Peak height at tide gauge,
in centimeters above
ambient tide
- ☐ 54

Kuwagasaki
100
38
Ōtsuchi
42,92,54
84
Nakaminato
33   47   54
20   83   Miho
Tanabe
98
25
54
54

1964 ALASKA TSUNAMI: MINIMAL IN JAPAN

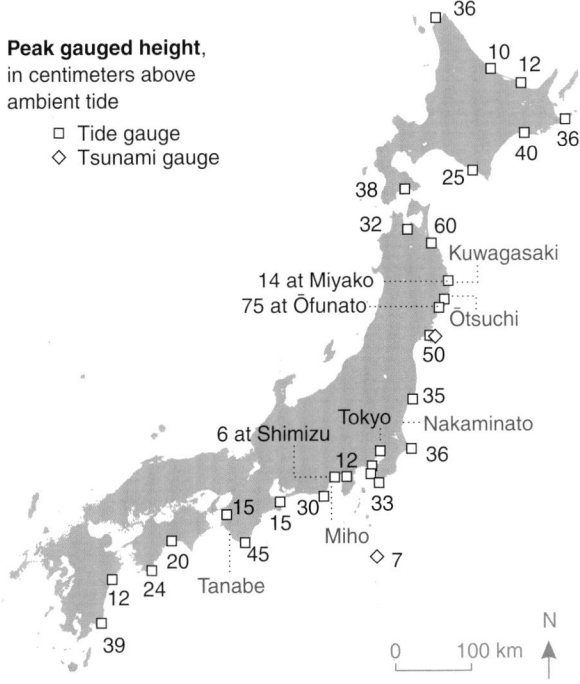

Peak gauged height,
in centimeters above
ambient tide
- ☐ Tide gauge
- ◇ Tsunami gauge

36
10   12
40   36
25
38
32   60
Kuwagasaki
14 at Miyako
75 at Ōfunato
Ōtsuchi
50
35
Tokyo
6 at Shimizu   Nakaminato
12   36
15   30   33
15   Miho
20   45   ◇ 7
12   24   Tanabe
39

0   100 km   N

FOR FURTHER CONTRAST with the 1960 Chile tsunami, compare both these maps with the ones on page 55. The Kamchatka tsunami heights are from The Central Meteorological Observatory (1953, p. 39, 45-58); the Alaskan data, from Hatori (1965).

TP, a vertical datum near mean sea level.

Alaskan ancestors

EVIDENCE AGAINST an Alaskan source for the 1700 tsunami includes not just the modest size of the 1964 Alaska tsunami in Japan but also the geologic history of pre-1964 Alaska earthquakes.

The immediate predecessor of the 1964 Alaska earthquake predates 1700 by 400 years or more. At upper Cook Inlet, where a buried soil marks land subsidence from 1964 (p. 14-15), an underlying buried soil dates the penultimate subsidence event to A.D. 1000-1200 (below). Similarly at the Copper River delta, uplifted in 1964, the penultimate uplift occurred about 1100-1300.

150°W
Cook Inlet
60°N
Girdwood
Copper River delta
Approximate rupture area
of 1964 Alaska earthquake
0   500 km   N

Girdwood junction
Sitka spruce killed in 1964
Bank in photo below
1991

A.D. 1964   1000-1200
2003

ON PREDECESSORS to the 1964 Alaska earthquake, see Combellick (1991), Plafker and others (1992), and Hamilton and Shennan (2005).

THE PHOTOS show the shore of Turnagain Arm at Girdwood. Lower image courtesy of Ian Shennan.

Tree-ring tests 年輪のテスト

A great Cascadia earthquake killed red-cedar trees between August 1699 and May 1700.

IN 1996, soon after Japanese researchers assigned a Cascadia earthquake to January 1700, North Americans sought to test the date. Radiocarbon had already been pushed to its limits in dating the death of earthquake-killed trees as exactly as 1695-1720 (p. 24-25). But there remained the possibility of dating, to the year and growing season, the trees' final months of growth.

That work had begun in 1987 with sampling of the red-cedar trunks standing in tidal wetlands of four Washington estuaries (photos, p. 16, 24; red diamonds, right). The victims contain a climatic bar code: year-to-year variation in the width of their annual rings. They share the code with old trees that safely witnessed the earthquake from high ground (cartoon, opposite). Witnesses felled by loggers in 1987 give the year for each bar in the code. Matching of the ring-width patterns thus yields dates for the victims' rings.

Dating a victims' year of death, however, requires samples that preserve the tree's final ring. The samples dated in the 1980s came instead from weather-beaten trunks. So in the summer of 1996, to ask trees whether they died from an earthquake in January 1700, geologists unearthed bark-bearing roots attached to the already-dated trunks. Tree-ring scientists then checked the ring-pattern match between root and trunk. The work yielded, for each of eight trees, a final-ring date. In all but one case, the tree died after completing the 1699 growing season and before the start of the next—in the window between August 1699 and May 1700.

As a further test, tree-ring scientists dated the onset of stress in Sitka spruce that barely survived post-earthquake tides (yellow triangles). The trees endured the submergence by sprouting roots into the new, higher ground. Several dozen such survivors remained in southern Washington and northern Oregon in the early 1990s. In half of them the width or anatomy of annual rings changed in 1700-1710 (examples in box, opposite).

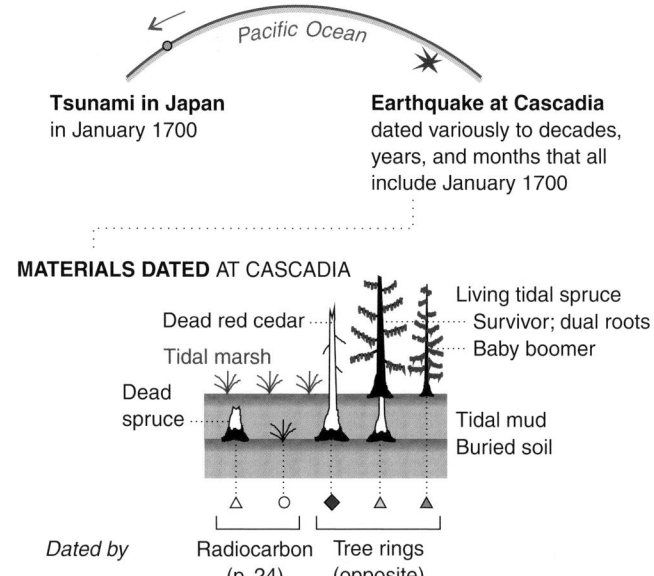

Tsunami in Japan
in January 1700

Earthquake at Cascadia
dated variously to decades, years, and months that all include January 1700

MATERIALS DATED AT CASCADIA

Dead red cedar
Tidal marsh
Dead spruce
Living tidal spruce
Survivor; dual roots
Baby boomer
Tidal mud
Buried soil

Dated by Radiocarbon (p. 24) Tree rings (opposite)

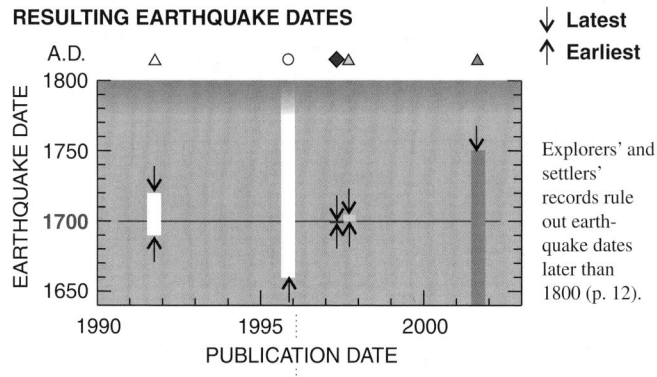

RESULTING EARTHQUAKE DATES

↓ **Latest**
↑ **Earliest**

A.D.

EARTHQUAKE DATE

PUBLICATION DATE

Explorers' and settlers' records rule out earth-quake dates later than 1800 (p. 12).

In early 1996, a Cascadia earthquake was assigned to January 26, 1700 by linkage to a tsunami in the written history of Japan (p. 42, 94).

BEST-DATED SITES

RING-WIDTH PATTERNS were matched to date the ring next to bark in the roots of eight red cedar (Yamaguchi and others, 1997; Jacoby and others, 1997). Seven of these trees died between the 1699 and 1700 growing seasons; the other survived until 1708. The ring-width measurements from the trunks of witnesses and victims are archived at ftp://ftp.ncdc.noaa.gov/pub/data/paleo/treering as rwl files wa129 through wa133.

STRESS IN SURVIVING SPRUCE was documented by Jacoby and others (1997). Aside from a few dozen survivors, living spruce of Washington's tidal forests postdate 1700. Most of the trees postdate 1750 because of a lag in colonizing lands that brackish tides were rebuilding (Benson and others, 2001).

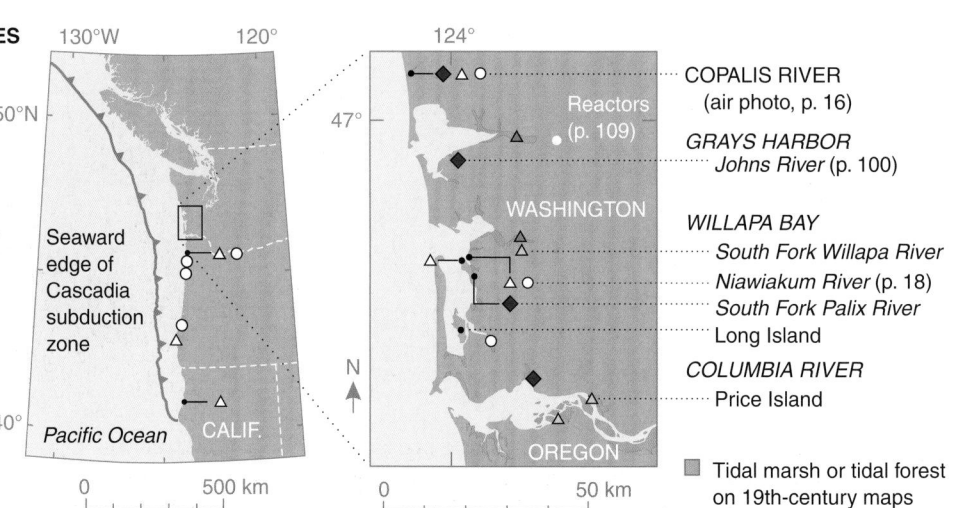

Seaward edge of Cascadia subduction zone

Pacific Ocean

CALIF.

COPALIS RIVER
(air photo, p. 16)

GRAYS HARBOR
Johns River (p. 100)

WILLAPA BAY
South Fork Willapa River
Niawiakum River (p. 18)
South Fork Palix River
Long Island

COLUMBIA RIVER
Price Island

Reactors (p. 109)

WASHINGTON

OREGON

▨ Tidal marsh or tidal forest on 19th-century maps

◆ Matched ring-width patterns of western red cedar

BAR-CODE ANALOGY

Witness on hill

A.D. 993   1700   1986

Victim beside bay

Final rings

— Eroded on exposed trunk

— Preserved in buried root

WITNESS'S INTACT TRUNK, WILLAPA BAY

A.D. 1690   1700   1710   1720   1730

Genroku era (p. 42, 63)   1 cm

One year's growth begins in spring and early summer with light-colored "early wood." The growing season concludes in late summer or early fall with dark "late wood."

In buried roots of red-cedar victims (example below), the 1699 ring contains both early wood and late wood—evidence that the trees lived through the 1699 growing season.

VICTIM'S WEATHER-BEATEN TRUNK, WILLAPA BAY

1603: Edo period begins (p. 37).   1661

1630

1 cm

Rough, weather-beaten exterior   Bark and tens of outer rings lost to centuries of wind, rain, decay, insects, birds, and fire.

INTACT ROOT OF THAT VICTIM (p. 92)

1630   1661   1699

1 mm

Smooth, unweathered exterior was covered with bark when chain-sawed in 1996.

△ Signs of stress in surviving Sitka spruce

RING WIDTHS OF TWO SURVIVORS

before | after January 1700   before | after

1 mm [

South Fork Willapa River

Price Island

A.D. 1700   1800   1900   1700   1720

GROWING SEASON

Price Island, 1994

SURVIVORS' GROVE, COLUMBIA RIVER

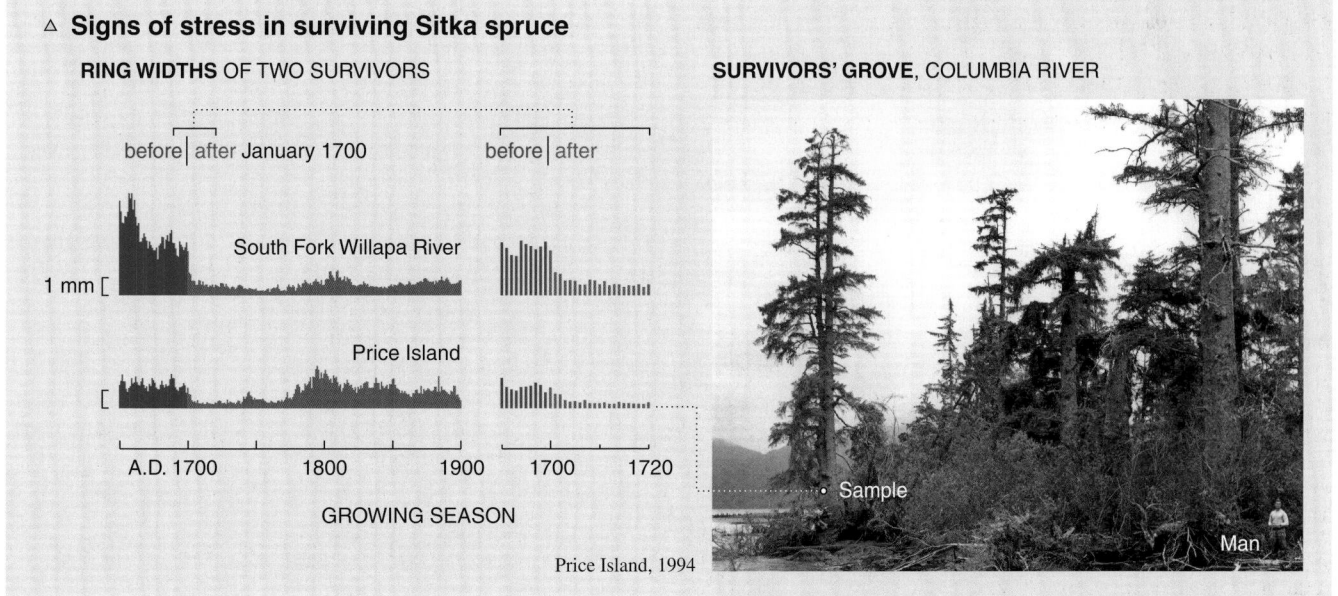

Sample

Man

ON TREE-RING DATING see Stokes and Smiley (1968), Fritts (1976), and Schweingruber (1988). **Witness** is red cedar from land above the reach of post-earthquake tides, at Long Island—from one of 19 used to make a master bar code for A.D. 993-1986 (Yamaguchi and others, 1997). **Victim** tree is PX-782, a stump along the South Fork Palix River (entire cross-section of root, p. 92). **Survivor** data is from Jacoby and others (1997).

Magnitude 9 マグニュード9

The 1700 Cascadia earthquake probably attained a magnitude between 8.7 and 9.2.

MAGNITUDE 8.7-9.2 COMPARED WITH MAGNITUDES OF GREAT EARTHQUAKES SINCE 1900

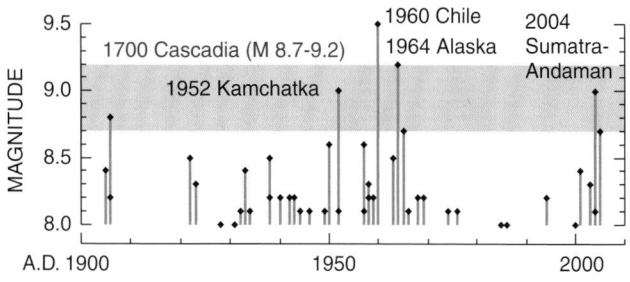

ENERGY RADIATED, ON LINEAR SCALE

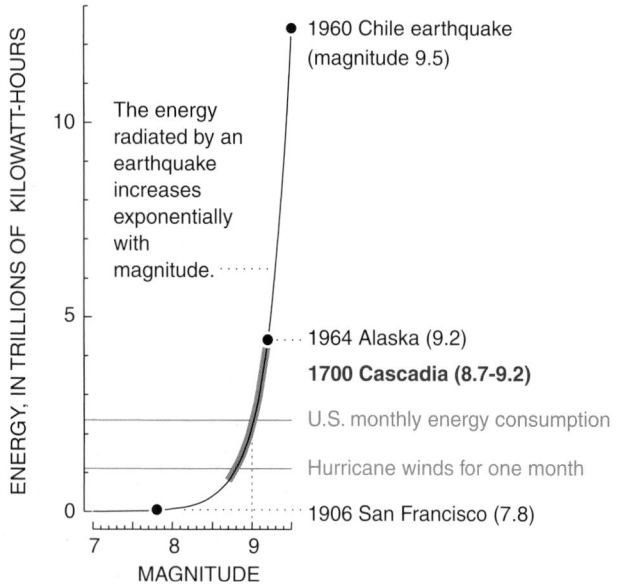

RUPTURE AREA COMPARED WITH HONSHU AND CALIFORNIA

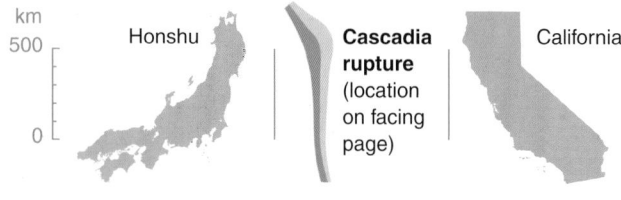

EARTHQUAKE MAGNITUDES from Kanamori (1977), Johnson and others (1994, p. 24), http://earthquake.usgs.gov/docs/sign_eqs.htm, Satake and others (2003), and Lay and others (2005). For comparison with 1960 Chile and 1964 Alaska, the most appropriate size of the 2004 Sumatra-Andaman earthquake is M 9.0 (footnote, p. 5). The linear energy scale is adapted from Johnson (1990).

U.S. ENERGY CONSUMPTION is for the year 2001 (www.eia.doe.gov/emeu/cabs/contents.html).

HURRICANE WIND ENERGY computed as 1.5 x 10^12 watts per day for a wind speed of 40 meters per second (90 miles per hour) in a radius of 60 km (www.aoml.noaa.gov/hrd/tcfaq/D7.html). The hurricane-force winds of Hurricane Isabel had about this combination of speed and area when the storm made landfall in North Carolina on September 18, 2003

A MAGNITUDE OF 9 makes an earthquake unusually enormous. Only two twentieth-century earthquakes surpassed M 9.0 (left). In several minutes, an earthquake of M 9.0 radiates as much energy as the United States consumes in a month, or twice the energy a hurricane's winds would release if they blew nonstop for a month (middle graph).

The 1700 Cascadia earthquake probably was such a giant. It likely broke at least 1,000 kilometers of the boundary between the subducting Juan de Fuca Plate and the overriding North America Plate—a rupture about as long as California, or about the length of Japan's main island, Honshu (lower left). On the seaward half of the rupture, the plates probably lurched past one another by about 20 meters. The magnitude was probably in the range M 8.7-9.2.

These estimates depend, in part, on assumptions about what fault area broke during the 1700 earthquake. By the assumptions in red at right, the break extends about 1,100 km coastwise and averages nearly 100 km in width. The fullest seismic slip takes place offshore, where the break is shallow (dark). Onshore the slip diminishes toward depths where the fault is too warm for brittle failure (light).

This picture has gained support from orbiting satellites of the Global Positioning System. GPS measurements help define mostly offshore areas where the downgoing Juan de Fuca Plate is currently coupled with the overriding North America Plate. Farther inland, the plates episodically creep a few centimeters past one another (green).

Resulting estimates of fault-rupture areas provide a starting point for simulating, by computer, the sea-floor displacement that triggered the 1700 tsunami. Offshore the sea floor rises several meters as the North America Plate lurches up the inclined fault. Near the coast, the seafloor and the adjacent land fall as much as two meters as this plate stretches (cartoons, p. 10). The simulated deformation varies with the rupture width and the slip amount—two of the main contributors to earthquake size.

Additional simulations track the resulting tsunami across the Pacific Ocean (p. 74-75). The modeled tsunami heights in Japan can then be compared with the heights estimated from damage and flooding by the orphan tsunami (bar graph, opposite). The comparisons rule out a Cascadia parent of M 8.0-8.5, whose tsunami would not likely exceed 1 m high in Japan. Instead, the inferred combinations of rupture area and seismic slip correspond to Cascadia earthquakes of M 8.7-9.2, with the best fit at M 9.0.

MAGNITUDE 8.7-9.2 explains three sets of reconstructed tsunami heights in Japan (p. 48), six assumed rupture areas at Cascadia, and various amounts of seismic slip in each of these rupture areas (Satake and others, 2003). The rupture depicted on the facing page is among three found consistent with geologic evidence for coastal subsidence like that on pages 16 and 17. The range M 8.7-9.2 excludes errors from ignoring bottom friction in computing the tsunami's advance through shallow water off Japan.

Rupture and deformation from a hypothetical 1700 earthquake

RUPTURE AREA **VERTICAL DISPLACEMENT**

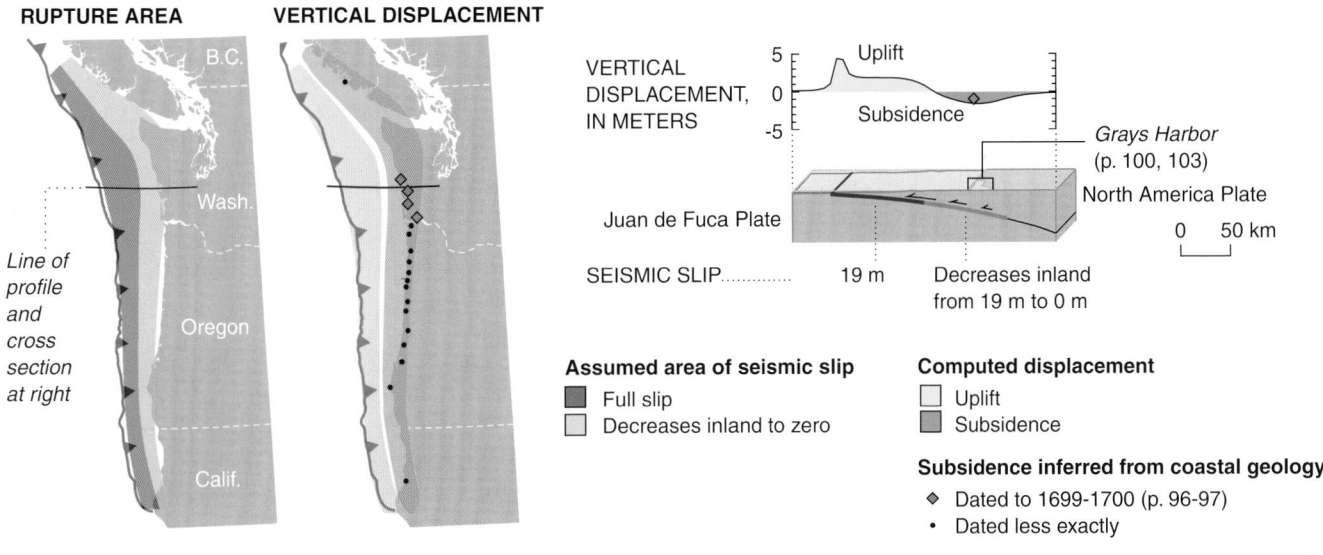

Line of profile and cross section at right

VERTICAL DISPLACEMENT, IN METERS

Uplift
Subsidence

Grays Harbor (p. 100, 103)

North America Plate

Juan de Fuca Plate

0   50 km

SEISMIC SLIP............. 19 m   Decreases inland from 19 m to 0 m

Assumed area of seismic slip
- Full slip
- Decreases inland to zero

Computed displacement
- Uplift
- Subsidence

Subsidence inferred from coastal geology
- ◆ Dated to 1699-1700 (p. 96-97)
- • Dated less exactly

Modern motions that help define the rupture area in 1700

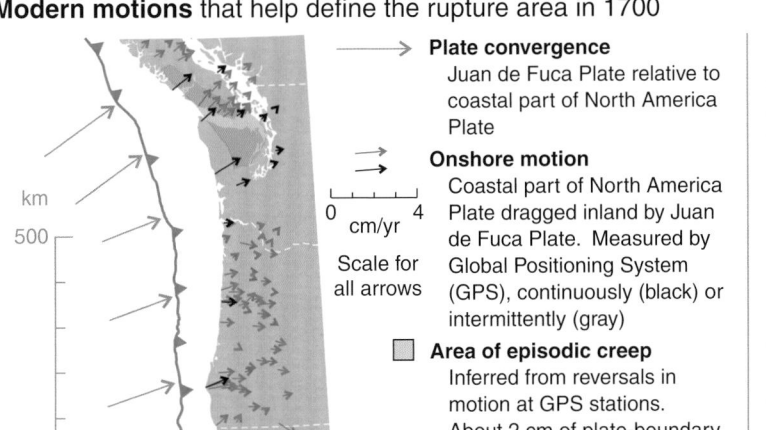

→ **Plate convergence**
Juan de Fuca Plate relative to coastal part of North America Plate

⇒ **Onshore motion**
Coastal part of North America Plate dragged inland by Juan de Fuca Plate. Measured by Global Positioning System (GPS), continuously (black) or intermittently (gray)

0   cm/yr   4
Scale for all arrows

▢ **Area of episodic creep**
Inferred from reversals in motion at GPS stations. About 2 cm of plate-boundary slip per event, at intervals of 13-16 months.

DAILY GPS DATA, SMOOTHED TREND

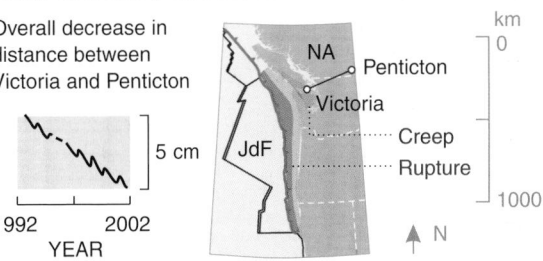

Overall decrease in distance between Victoria and Penticton

5 cm

1992   2002
YEAR

The North America Plate (NA) shortened between Victoria and Penticton by 5 cm between 1992 and 2002, probably from being stuck to the subducting Juan de Fuca Plate (JdF) in the area of the inferred 1700 rupture (red and pink). But the continent also extended on occasion, during episodes of creep on a deeper part of the plate boundary (green). The extension produces the sawteeth on the graph.

Modeled Japanese tsunami heights for the earthquake, compared with heights inferred from flooding and damage

TSUNAMI MODEL   **BEST-FIT COMPARISON**

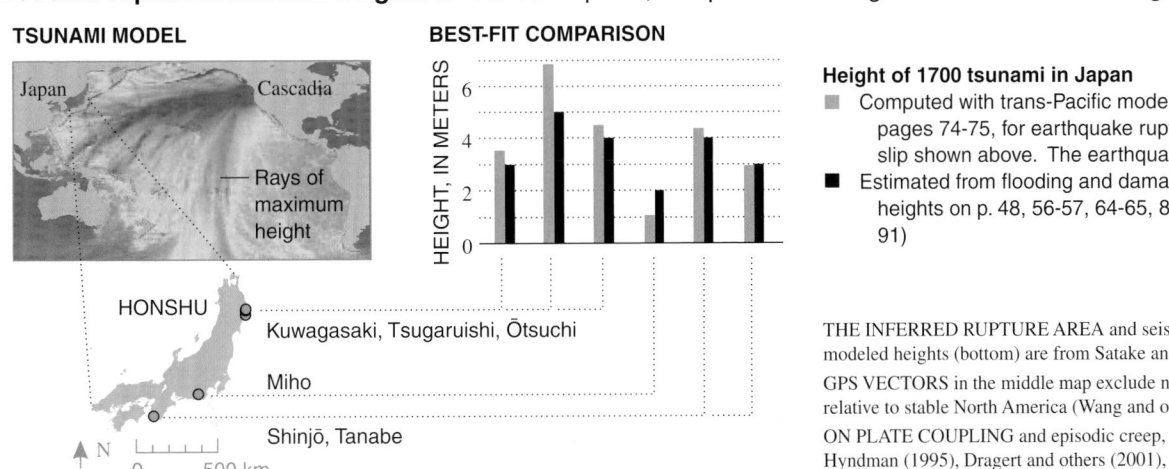

Rays of maximum height

HEIGHT, IN METERS

HONSHU

Kuwagasaki, Tsugaruishi, Ōtsuchi

Miho

Shinjō, Tanabe

N   0   500 km

Height of 1700 tsunami in Japan
- ▢ Computed with trans-Pacific model at left and on pages 74-75, for earthquake rupture area and slip shown above. The earthquake size is M 9.0.
- ■ Estimated from flooding and damage (the "B" heights on p. 48, 56-57, 64-65, 80-81, and 88-91)

THE INFERRED RUPTURE AREA and seismic slip (top) and modeled heights (bottom) are from Satake and others (2003).

GPS VECTORS in the middle map exclude northward motion relative to stable North America (Wang and others, 2003).

ON PLATE COUPLING and episodic creep, see Dragert and Hyndman (1995), Dragert and others (2001), Miller and others (2002), Rogers and Dragert (2003), Szeliga and others (2004), and Melbourne and others (2005).

Muddy forecast　泥から森へ

How will history repeat itself at Cascadia?

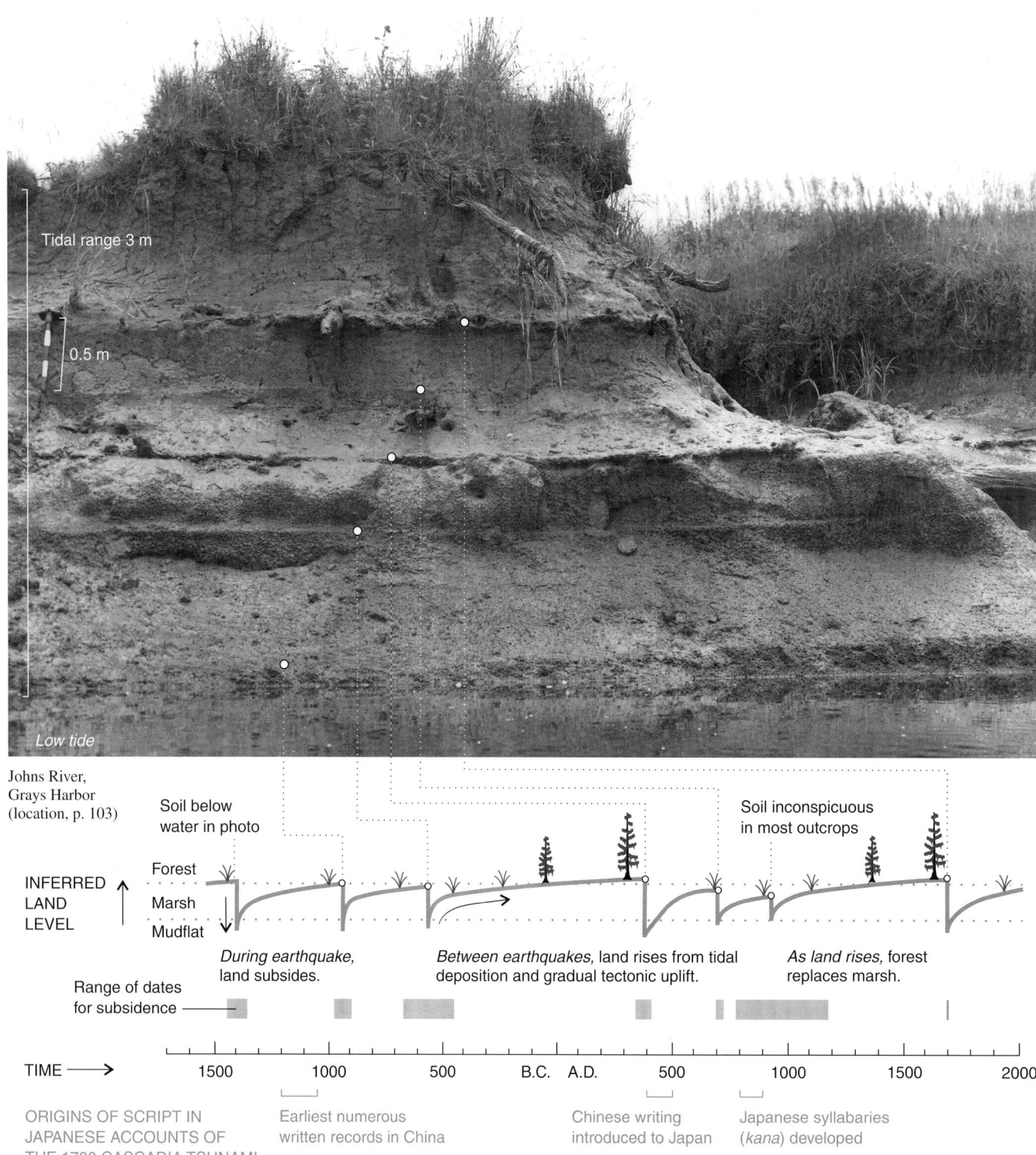

Tidal range 3 m

0.5 m

Low tide

Johns River,
Grays Harbor
(location, p. 103)

**Soil below
water in photo**

**Soil inconspicuous
in most outcrops**

INFERRED
LAND
LEVEL

Forest

Marsh

Mudflat

During earthquake,
land subsides.

Between earthquakes, land rises from tidal
deposition and gradual tectonic uplift.

As land rises, forest
replaces marsh.

Range of dates
for subsidence

TIME ⟶   1500   1000   500   B.C. A.D.   500   1000   1500   2000

ORIGINS OF SCRIPT IN
JAPANESE ACCOUNTS OF
THE 1700 CASCADIA TSUNAMI

Earliest numerous
written records in China

Chinese writing
introduced to Japan

Japanese syllabaries
(*kana*) developed

THE EARTHQUAKE TIMELINE applies to Grays Harbor, Willapa Bay, and the
Columbia River estuary (location map, p. 96). The gray bars span 95-percent
confidence intervals from radiocarbon dating reported by Atwater and others
(2004). The pictured outcrop adjoins site JR-1 of Shennan and others (1996).

ASIAN SCRIPTS in accounts of the 1700 tsunami evolved through at least five of

the intervals between great Cascadia earthquakes. Writings from China's Shang
dynasty—inscribed into cattle scapulas and turtle shells—date to 1200-1050 B.C.
(Keightley, 1978, p. 228). Early examples of Chinese characters written in Japan
date to the 5th century A.D. (Seeley, 2000, p. 4-6, 16-25). Japanese phonetic
symbols became commonplace by early in the 11th century (Seeley, 2000, p. 76).

THE NEXT GREAT CASCADIA EARTHQUAKE is inevitable. The Cascadia plate boundary has repeatedly broken in great earthquakes during past millenia (summary graphs, below). Since 1700 the fault has been accumulating strain that future earthquakes will release (p. 99).

That next earthquake may have already happened by the time you read this, or it may come lifetimes later. Cascadia makes earthquakes on an irregular schedule.

In the example of irregularity at left, a low-tide outcrop in Washington displays buried soils from each of five great earthquakes of the past 3,000 years. Another buried soil lies below low tide, and still another is too poorly preserved to form a visible ledge. The full sequence tells of seven earthquakes from the past 3,500 years. The seven intervals average about 500 years but range approximately from 200 years to 1,000 years. The two longest are marked by extensive remains of forests; the extra time allowed tidal land

to rise high enough to become much more widely forested than it is today (bottom of facing page).

During Cascadia's next great earthquake, will the plate boundary rupture along its full length, as in 1700, or will it break one piece at a time? Either behavior would be consistent with geologic records of great Cascadia earthquakes. Piecemeal rupture can't be ruled out (p. 24-25), especially if Cascadia behaves like subduction zones where successive earthquakes differ in size (box, below).

For now it is prudent to assume, simplistically, that the next great Cascadia earthquake has a one-in-ten chance of occurring in the next 50 years, and that it may attain magnitude 9 (p. 102-105). The one-in-ten odds follow from an average interval of 500 years if the fault lacks memory of when it last broke. The magnitude-9 assumption leaves a margin of safety in case of lesser events.

AVERAGE INTERVALS BETWEEN GREAT CASCADIA EARTHQUAKES

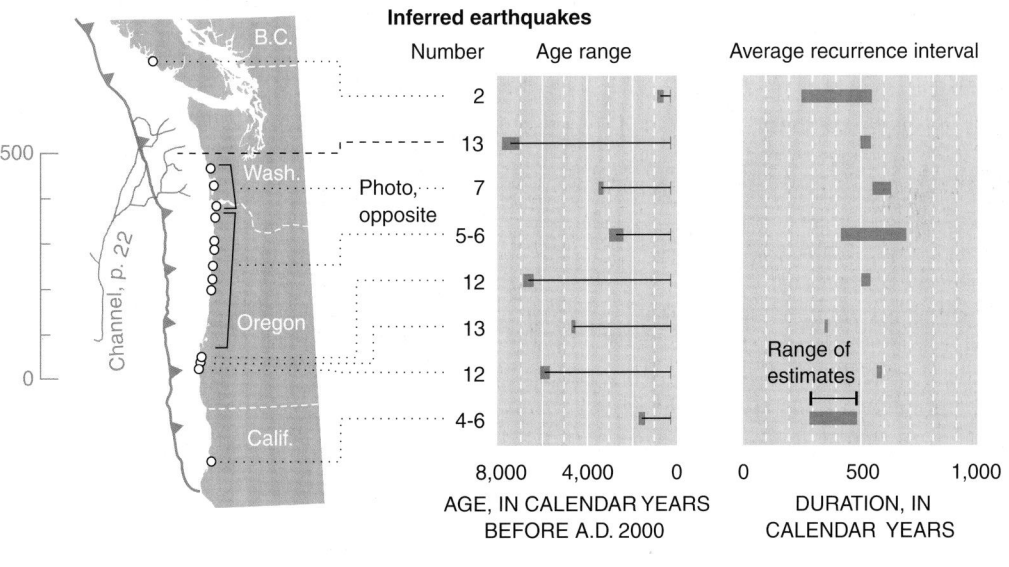

SOURCE OF ESTIMATES

Clague and Bobrowsky (1994)

Adams (1990)

Atwater and others (2004)

Darienzo and Peterson (1995)

Witter and others (2003)

Kelsey and others (2005)

Kelsey and others (2002)

Clarke and Carver (1992)

Peterson and others (2002) and Mazzotti and Adams (2004) derive conditional probabilities of great Cascadia earthquakes.

Versatile faults

A SUBDUCTION ZONE that breaks in a long rupture may also rupture in shorter pieces. At Japan's Nankai Trough, the rupture area of a single earthquake in 1707 slipped next in a pair of lesser earthquakes in 1854 and again in two parts in the 1940s (map, p. 85). Similarly in South America and South Asia, single earthquake ruptures have spanned the areas of multiple, smaller breaks. Variable rupture can be expected at Cascadia as well.

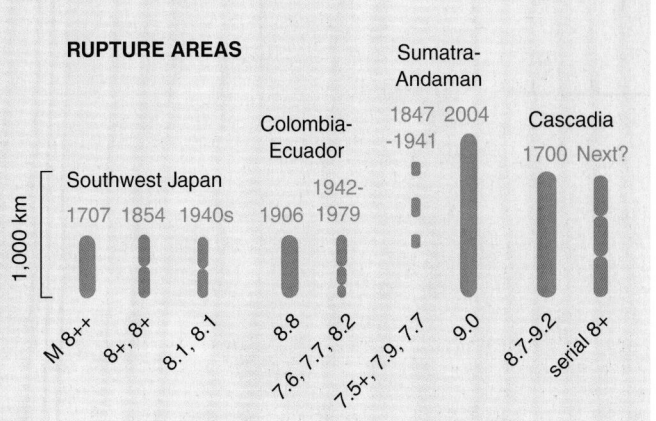

Japan, Ando (1975); Colombia-Ecuador, Kanamori and McNally (1982); Sumatra-Andaman, Bilham and others (2005).

High-enough ground 安全な高さとは？

What places offer refuge from a Cascadia tsunami?

PLANS FOR FLEEING TSUNAMIS in North America have been shaped by the Japanese accounts of the 1700 tsunami. The accounts, along with Native American traditions, have spurred such planning by providing eyewitness evidence for a giant Cascadia tsunami. Moreover, because the Japanese accounts suggest a Cascadia earthquake of magnitude 9, they provide a basis for evacuation signs and maps, such as those at right. Since 1997, tsunami mapping at Cascadia has been based on computer modeling of a Cascadia earthquake of M 9.1. The modelers chose this magnitude to resemble the one inferred, in 1996, from heights of the 1700 tsunami in Japan.

Since 1997, tsunami modeling has identified inundation-prone areas in cities and towns along Washington's outer coast and on parts of the Oregon coast (index map, facing page). Evacuation maps based on the modeling serve most of the U.S. mainland population at risk from a great Cascadia tsunami. That at-risk population exceeded 150,000 year-round residents in the year 2000, as judged from census totals for areas within 1 km (0.6 mi) of tidewater.

The tsunami mapping helps citizens and public officials identify areas of probable danger and of probable safety. The evacuation map for Gearhart, for example, shows where to assemble on high ground. The inundation map for Grays Harbor, opposite, similarly identifies a likely island of safety above a simulated tsunami in Westport. Farther inland at Aberdeen, the map depicts inundation that could turn logs into battering rams.

The models fit geologic evidence for the 1700 tsunami. The areas of computed inundation commonly contain sand sheets from the flooding in 1700. Sequences of computed water levels, such as those graphed opposite, show multiple waves like those recorded by tide gauges (p. 19, 49, 73) and by sediment layers (p. 18-19).

In simulations, the model tsunami has the advantage of overrunning freshly subsided land—land lowered as much as 1.5 meters (5 feet) during the parent earthquake. This is the subsidence anticipated on page 10, inferred from geology on pages 16-17, dated to 1700 or thereabouts on pages 24-25 and 96-97, and computed for a model rupture on page 99. The coast's subsidence during an earthquake increases the hazard from the ensuing tsunami.

ROADSIDE ADVICE IN HOQUIAM, WASHINGTON

2003

TSUNAMI EVACUATION MAP FOR GEARHART, OREGON

Evacuation area   Upper limit 12 m (40 ft) near beach, 6 m (20 ft) farther inland
⇒ Evacuation route
Ⓐ Assembly area
Ⓕ Fire and police station

1 km
0.5 mi

THE FIRST MAPS of hazards from a Cascadia tsunami showed potential inundation in northern California. They were based on a computer model in which a hypothetical wave is 10 m high in offshore waters 50 m deep (Bernard and others, 1994; Toppozada and others, 1995).

OREGON'S LEGISLATURE soon mandated tsunami-inundation mapping of their entire coast. Under Senate Bill 379, passed in 1995 and implemented as Oregon Revised Statutes 455.446 and 455.447, new schools, hospitals, fire stations, and police stations shall not be constructed in areas subject to flooding by tsunamis, except where no alternative sites exist (http://www.leg.state.or.us/ors/455.html).

EVACUATION MAPS cover the Oregon towns of Bandon, Brookings, Charleston, Coos Bay, Depoe Bay, Gearhart (above), Gold Beach, Lincoln Beach and vicinity, Manzanita, Nehalem, Nestucca, Netarts, Newport, Oceanside, Port Orford, Rockaway Beach, and Seaside (http://sarvis.dogami.state.or.us/earthquakes/coastal/tsubrochures.htm), and the Washington communities of Aberdeen, Bay Center, Clallam Bay, Copalis, Cosmopolis, Hoquiam, Ilwaco, Long Beach, Neah Bay, North Cove, Ocean City, Ocean Park, Ocean Shores, Pacific Beach, Port Angeles, Port Townsend, Quilleyute, Raymond, Sound Bend, and Westport (http://www.dnr.wa.gov/geology/hazards/tsunami/evac/; http://emd.wa.gov/5-prog/prgms/eq-tsunami/tsunami-idx.htm). Locations in index (p. 125-133).

ELEMENTS OF TSUNAMI RISK FOR A CASCADIA EARTHQUAKE LIKE THAT OF 1700

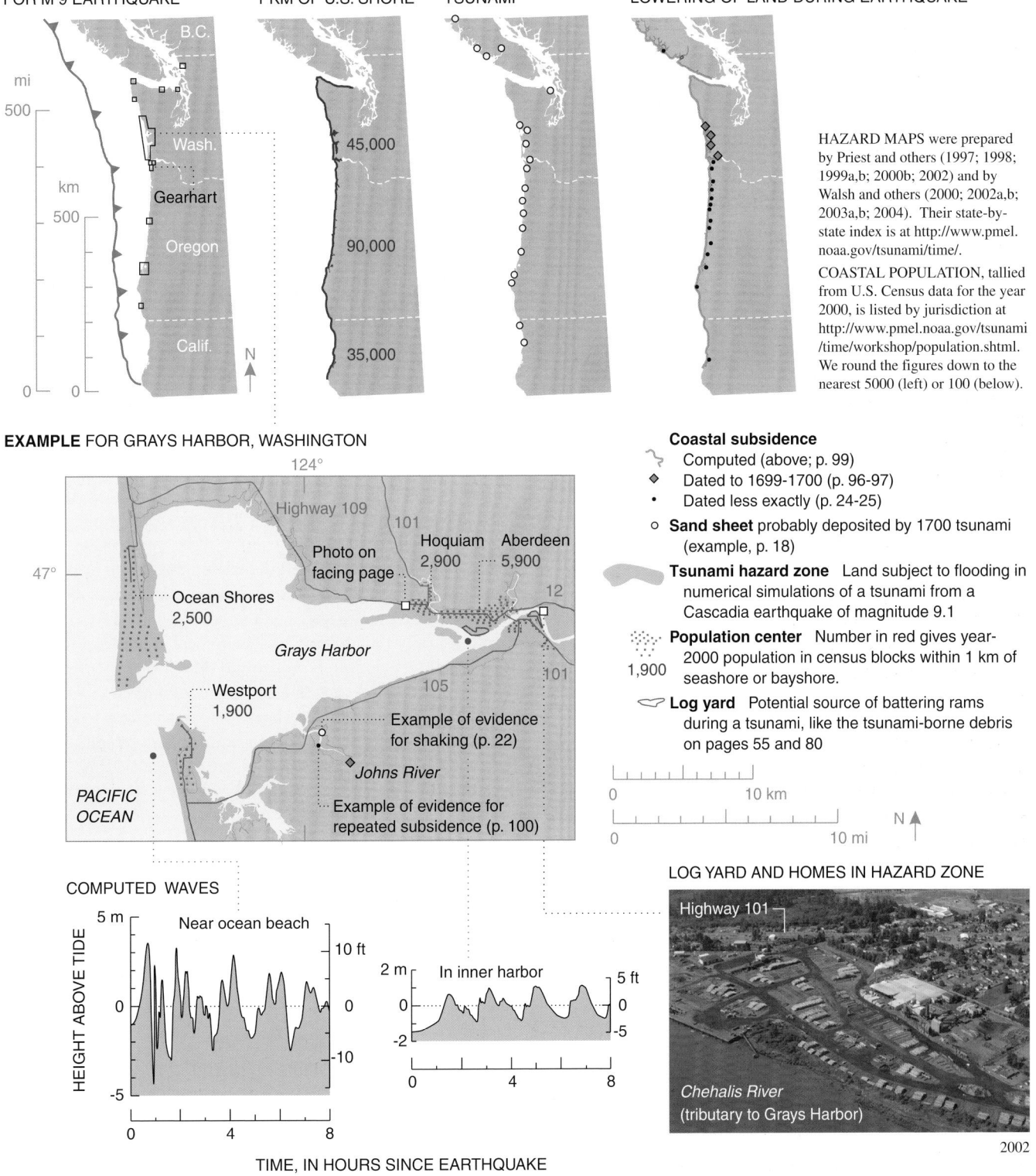

TSUNAMI HAZARD MAPPED FOR M 9 EARTHQUAKE

POPULATION WITHIN 1 KM OF U.S. SHORE

45,000

90,000

35,000

EVIDENCE FOR 1700 TSUNAMI

HAZARD POTENTIALLY INCREASED BY LOWERING OF LAND DURING EARTHQUAKE

HAZARD MAPS were prepared by Priest and others (1997; 1998; 1999a,b; 2000b; 2002) and by Walsh and others (2000; 2002a,b; 2003a,b; 2004). Their state-by-state index is at http://www.pmel. noaa.gov/tsunami/time/.

COASTAL POPULATION, tallied from U.S. Census data for the year 2000, is listed by jurisdiction at http://www.pmel.noaa.gov/tsunami /time/workshop/population.shtml. We round the figures down to the nearest 5000 (left) or 100 (below).

EXAMPLE FOR GRAYS HARBOR, WASHINGTON

Coastal subsidence
- Computed (above; p. 99)
- Dated to 1699-1700 (p. 96-97)
- Dated less exactly (p. 24-25)
- **Sand sheet** probably deposited by 1700 tsunami (example, p. 18)

Tsunami hazard zone Land subject to flooding in numerical simulations of a tsunami from a Cascadia earthquake of magnitude 9.1

Population center Number in red gives year-2000 population in census blocks within 1 km of seashore or bayshore.

Log yard Potential source of battering rams during a tsunami, like the tsunami-borne debris on pages 55 and 80

COMPUTED WAVES

LOG YARD AND HOMES IN HAZARD ZONE

GRAYS HARBOR HAZARD MAP and wave-train simulations, from Walsh and others (2000), are based on computer modeling of an assumed earthquake rupture 1,050 km long and, on average, 70 km wide (Myers and others, 1999; Priest and others, 2000a). The seismic slip is uniform along the length of this hypothetical rupture. Tide stage is held steady near mean tide level. Not depicted is the

slightly greater tsunami modeled for a rupture that includes a patch of greater-than-average slip off Washington (asperity model of Walsh and others, 2000).

PHOTO from Washington Department of Ecology digital coastal atlas (http://apps.ecy.wa.gov/website/coastal_atlas/viewer.htm), image 0208081033_378.

The orphan's parent

Seismic waves 地震動

Tall buildings await Cascadia's next great earthquake.

SHORT, TRADITIONAL BUILDINGS

Seattle, 1884

TALL URBAN BUILDINGS

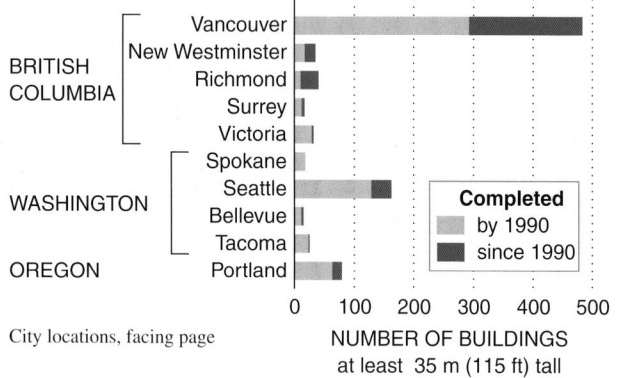

BRITISH COLUMBIA	Vancouver
	New Westminster
	Richmond
	Surrey
	Victoria
WASHINGTON	Spokane
	Seattle
	Bellevue
	Tacoma
OREGON	Portland

Completed
by 1990
since 1990

0   100   200   300   400   500

City locations, facing page

NUMBER OF BUILDINGS
at least 35 m (115 ft) tall

EARTHQUAKE SOURCES

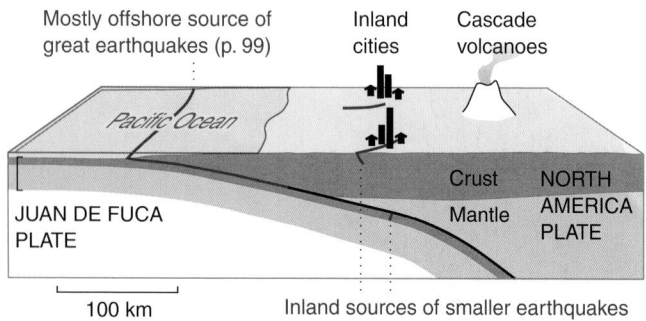

Mostly offshore source of great earthquakes (p. 99)

Inland cities

Cascade volcanoes

Pacific Ocean

Crust

Mantle

NORTH AMERICA PLATE

JUAN DE FUCA PLATE

100 km

Inland sources of smaller earthquakes

THE URBAN CORRIDOR between Vancouver, British Columbia, and Eugene, Oregon, can expect minutes of shaking from a great Cascadia earthquake. The shaking poses less of a threat to the region's traditional wood-framed houses than to larger structures that are slender and flexible. Tall buildings, long bridges, and steel aqueducts sway most readily at periods of a second or more. Great earthquakes excel in exciting such long-period motion. A common result, seen in 1985 in Mexico City, is damaging resonance between the ground and the long-period structures founded on it.

Despite its inland location, the urban corridor from Vancouver to Eugene lies within range of damaging ground motions from great Cascadia earthquakes. Long-period waves from subduction earthquakes can travel hundreds of kilometers inland without losing much of their punch. In addition, the waves can get trapped and amplified in sedimentary basins like those beneath Seattle and Tacoma.

Only recently did these threats become certain enough to affect building design. Among Cascadia's nearly 900 high-rises, more than half were completed by 1990 (graph). Not until 1994 did building codes in Washington and Oregon begin to reflect the great-earthquake threat. Even then, designers of newer structures faced a moving target as the credible size of a Cascadia earthquake rose to M 9 (p. 98-99), and as newly found urban faults augmented the hazard (block diagram).

The prospect of great Cascadia earthquakes influences the mapping of earthquake hazards in the western United States, especially for ground motions of long period. According to the maps at right, plate-boundary ruptures at Cascadia contribute to the hazard of long-period seismic shaking across Washington, Oregon, and northern California, particularly in the western parts of those states.

TALL BUILDINGS SWAY at fundamental periods of 1-6 seconds (Building Seismic Safety Council, 2001, p. 106). The 1985 Michoacan earthquake of M 8.0 caused inordinate damage to Mexico City high rises with fundamental periods of 1 second, at a distance 400 km from this subduction earthquake's source (p. 9; see also Scawthorn and Celebi, 1987).

BY LASTING A MINUTE OR MORE, a great Cascadia earthquake would likely cause more damage than would shaking of similar strength in a briefer earthquake (Tremblay, 1998).

STRONG SHAKING has been measured for earthquakes up to M 8.3 (Atkinson and Boore, 2003); ground motions for M 9 are extrapolations (Heaton and Hartzell, 1989). Beneath Seattle, a sedimentary basin several kilometers deep amplifies weak shaking at periods of 1-5 seconds (Pratt and others, 2003).

THE UNIFORM BUILDING CODE extended its seismic zone 3, for high hazard, throughout western Washington and western Oregon in 1994 (Atwater and others, 1995).

FOR ADDITIONAL INFORMATION on ground-shaking hazards at Cascadia, see Yeats (2004), Ballantyne and others (2005), and Cascadia Region Earthquake Workgroup (2005).

OLD SEATTLE BUILDINGS from lithograph at University of Washington Libraries, Special Collections, UW347. Tall-building tallies from http://www.emporis.info/en/. In the block diagram, geologic boundaries are based on interpretations by Parsons and others (1998) and Brocher and others (2003).

ON INLAND EARTHQUAKE SOURCES in the North America crust, see Bucknam and others (1992), Johnson and others (2001), Nelson and others (2003), and Sherrod and others (2004). On earthquakes within the underlying Juan de Fuca Plate, see Frankel and others (2002a) and Atkinson and Casey (2003).

MAIN SOURCE OF SEISMIC WAVES OF 1 CYCLE PER SECOND

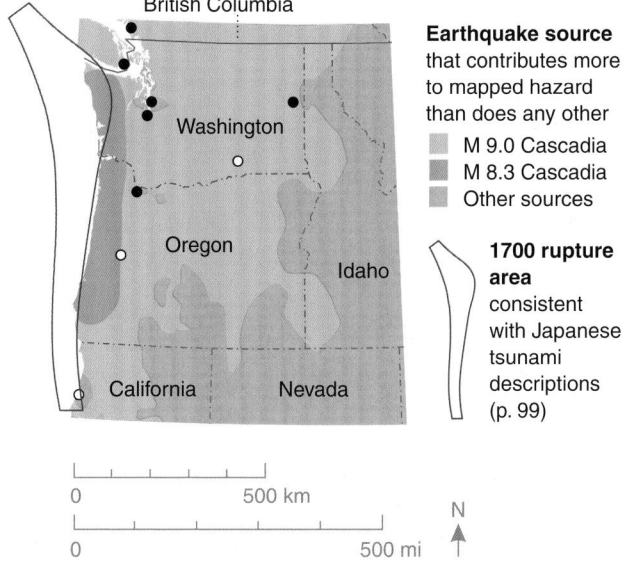

Earthquake source
that contributes more
to mapped hazard
than does any other

- M 9.0 Cascadia
- M 8.3 Cascadia
- Other sources

1700 rupture area
consistent
with Japanese
tsunami
descriptions
(p. 99)

0   500 km

0   500 mi

N ↑

GREAT-EARTHQUAKE CONTRIBUTION TO 1-SECOND HAZARD

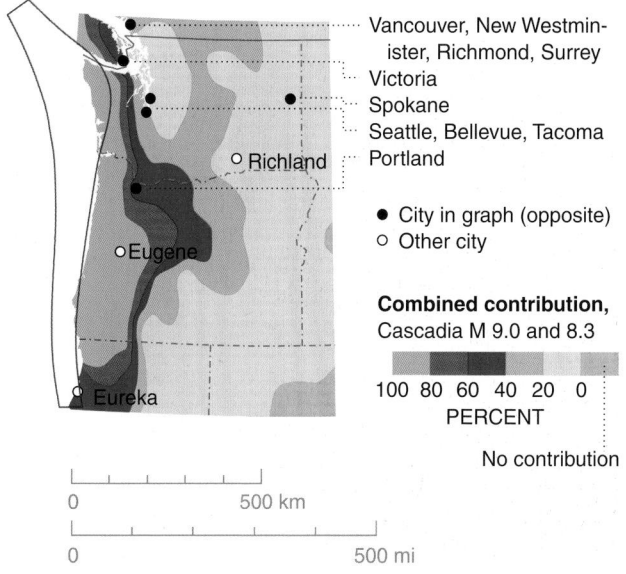

Vancouver, New Westmin-
ister, Richmond, Surrey
Victoria
Spokane
Seattle, Bellevue, Tacoma
Portland

● City in graph (opposite)
○ Other city

Combined contribution,
Cascadia M 9.0 and 8.3

100 80 60 40 20 0
PERCENT

No contribution

0   500 km

0   500 mi

THE MAPS ABOVE are derived from the 2002 version of national maps of earth-quake shaking hazards in the United States (Frankel and others, 2002b; maps at http://eqhazmaps.usgs.gov/). The national maps show the combined effect of hundreds of earthquake sources (such as the sources cartooned opposite). Companion maps and graphs deaggregate the shaking to show contributions from the individual sources (Harmsen and others, 2003). The deaggregations above were provided by Stephen C. Harmsen.

ASSUMPTIONS ABOUT CASCADIA earthquakes built into the national maps:
- Either the earthquakes attain M 9 with ruptures about 1,000 km long, or they are limited to M 8.3 ruptures 250 km long. These end-member scenarios were introduced in previous national hazard maps (Frankel and others, 1996, p. 16-17).
- M 9 is as plausible as M 8.3. "For 2002, we assigned a weight of 0.5 for each scenario... For 1996, the weights were 0.67 for the M 8.3 scenario and 0.33 for

the M 9.0 scenario. Since 1996, the M 9.0 scenario has gained credibility" (Frankel, and others, 2002b, p. 11).
- At a given place along the subduction zone, the mean recurrence interval for great earthquakes (either M 8.3 or M 9) is 520 years and the median interval 440 years. If the probability of earthquake occurrence does not vary with time within a recurrence interval, the resulting probability of either kind of event is about 10 percent in 50 years (Peterson and others, 2002, p. 2163-2164).

IN CANADA, great Cascadia earthquakes contribute to the hazard mapped for the proposed 2005 edition of the national building code. The assumed earthquake size is M 8.2, on the premise that only a nearby part of an M 9 rupture, comparable in size to a M 8.2 rupture, governs a site's shaking hazard (Adams and Atkinson, 2003, p. 260).

Shaking-hazard maps of the United States and Canada, including those above, do not yet reflect the long duration expected of great Cascadia earthquakes. An earthquake of M 9 would last several times longer than the largest earthquake expected of inland faults in the urban corridor. Engineers are beginning to grapple with how to design for shaking so prolonged.

It was a lack of seismic shaking that perplexed the Miho headman as he contemplated the orphan tsunami of 1700 (p. 54, 78-79). He or a later compiler recommended keeping the event in mind (right). Today, solved by geologic links to a distant earthquake, the headman's puzzle serves as a reminder to guard against infrequent earthquakes and tsunamis of extraordinary size.

nochi ni
Future

oite
in

kokoroe
keep in mind

tamau-beshi
should.

"Miho-mura yōji oboe," p. 78, columns 15-16.

Acknowledgments 謝辞

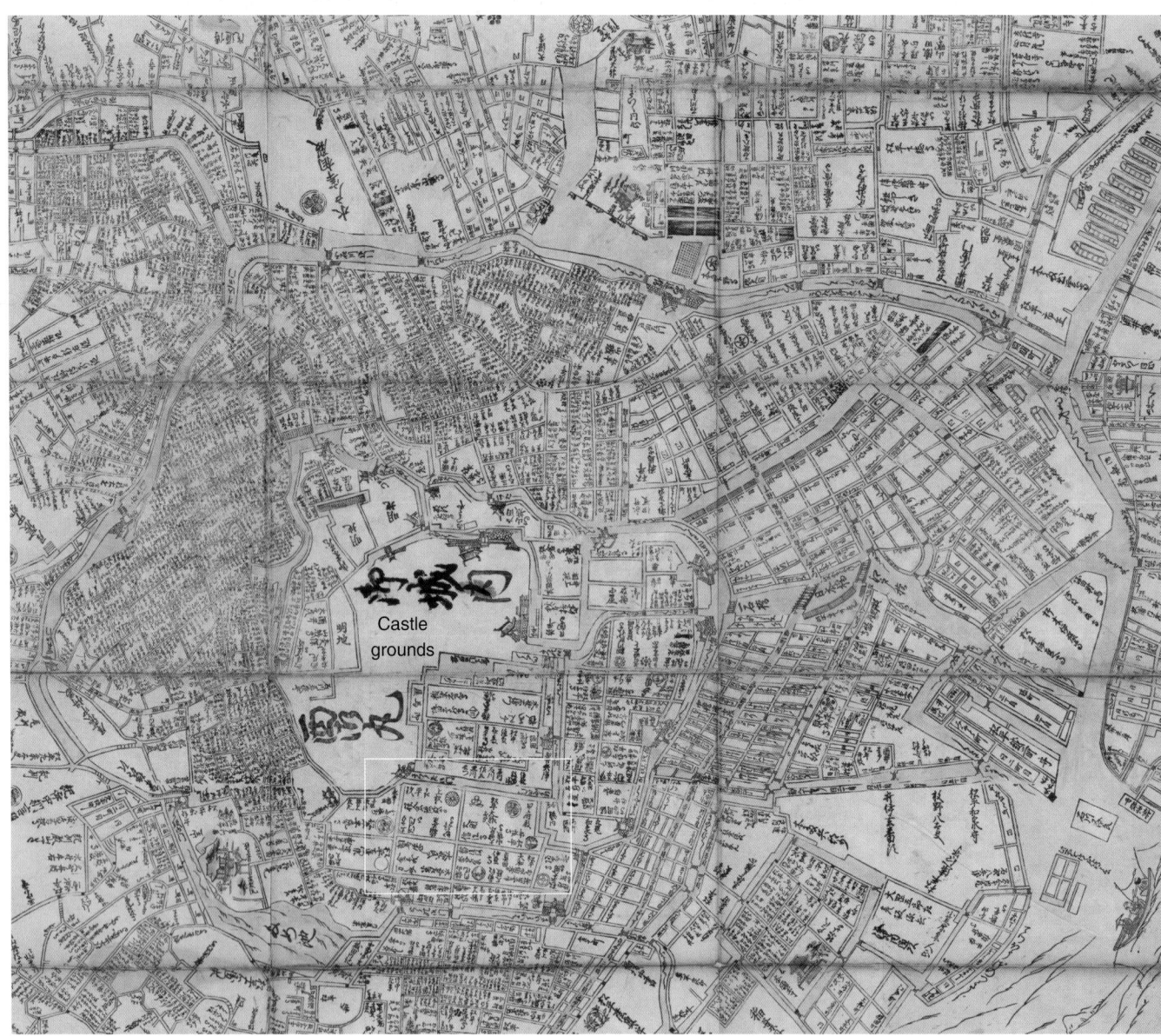

Castle grounds

Libraries in Berkeley, Morioka, Seattle, Tanabe, and Tokyo made available, for use in this book, maps that aid in visualizing the bygone world of the orphan tsunami of 1700. The example above—from the collection of the East Asian Library of the University of California, Berkeley—shows moats and samurai neighborhoods spiraling around the castle grounds of Edo in 1684. The white box outlines an area of daimyo mansions (enlarged view, p. 61).

1 km

↑ N

The Berkeley collection can be viewed at http://www.davidrumsey.com/japan/. The image above is excerpted from "Eiri Edo ōezu," published in Ten'na 4 by Hyōshiya Ichirōbe'e. Courtesy of East Asian Library, University of California, Berkeley.

WHERE JAPANESE WRITERS recorded the 1700 tsunami, dozens of people helped us explore questions central to this book: Who wrote the original accounts of the flooding and damage? Why were these accounts written and how were they preserved? Which passages contain errors in copying? Where are the places described as flooded? Were these same places reached by the 1960 tsunami as well?

In Morioka, Konishi Hiroaki granted access to the Morioka-han documents reproduced on pages 36, 38-39, 44-45, 58, and 60. He provided clues on how Morioka-han "Zassho" was compiled, documentation on senior ministers named there, and likely dates for the early 18th-century maps of Miyako-dōri and Ōtsuchi-dōri (p. 36, 44, 58). He serves as librarian of the Documents Office, Morioka City Central Community Center (Kyōdo Shiryō Shitsu, Morioka-shi Chūō Kōminkan).

On the coast in modern Miyako city, Yamazaki Toshio and Sasaki Tsutomu identified places inundated by the 1960 tsunami in Kuwagasaki and Tsugaruishi (photos, p. 49 and 51). In 1999, Mr. Yamazaki was fire chief and Mr. Sasaki one of his deputies at the Central Fire Station of the Miyako Unified Fire District (Miyako-chiku Kōiki Kumiai Gyōsei, Shōbōsho Honbu). Shuto Nobuo of Iwate Prefectural University provided an introduction to Mr. Yamazaki and a walking tour of Kuwagasaki's tsunami-prone districts. Kishi Shōichi, a historian for Miyako city, shared his knowledge of Miyako's Edo-period governance. His successor, Kariya Yūichirō, helped us interpret and photograph Moriai-ke "Nikki kakitome chō."

In Tsugaruishi, Moriai Mitsunori granted access to his family's notebook, Moriai-ke "Nikki kakitome chō." He and his mother welcomed three of us into the family home (p. 53). Iwamoto Yoshiteru, an authority on the area's Edo-period economy (books, p. 116), provided guidance on obscure place names of Tsugaruishi (p. 50, 51, 56).

Morikoshi Ryō of Hachinohe helped Ueda identify copyist's errors in Moriai-ke "Nikki kakitome chō" by providing a transcription, in printed Japanese, of official records of Hachinohe-han, its "Han nikki" (footnoted, p. 52). Mr. Morikoshi leads Hachinohe Komonjo Benkyō-kai, a group that studies historical documents and which made the transcription of Hachinohe "Han nikki."

Moriai Mutsuharu, a retired schoolteacher in Tsugaruishi, adopted Atwater and Yamaguchi for a day of interviewing his fellow villagers about the 1960 tsunami (sites marked by blue and yellow dots, p. 56). Those who identified inundation limits include Yonezawa Takuji (in color photo, p. 57, upper right) and Moriai Miya (photo, below).

In Ōtsuchi, Maeda Zenji, Fujimoto Toshiaki, and Kamata Seizō provided guidance on Edo-period neighborhoods. They also shared the town's collection of photographs and maps showing sites inundated by the 1960 tsunami. When interviewed in 1999, Mr. Maeda headed Ōtsuchi's Historical Preservation Council (Ōtsuchi-chō Bunkazai Hogo Shingikai), while Messrs. Fujimoto and Kamata served as assistant director and archaeologist, respectively, in the town's office of continuing education (Ōtsuchi-chō Kyōiku I'inkai, Shakai Kyōikuka).

Ogawa Kaori journeyed to Ōfunato to learn about that city's devastation by the 1960 tsunami and its lack of writings on the 1700 tsunami (p. 81). She also checked for written records in Sendai. In Ōfunato she received help from Satō Etsuro of Ōfunato city, Shirato Yutaka and Kin'no Ryōichi of Ōfunato Museum, and Honda Fumito of nearby Rikuzentakada city.

Town officials, local historians, and private citizens of Hitachinaka (formerly Nakaminato) twice received visitors interested in tsunami evidence from Ōuchi-ke "Go-yōdome." The hosts included Kawasaki Osamu, Onizawa Yōichi, Onizawa Yasuhiko, Saitō Arata, Satō Tsugio, and, from the family that conserves the document, Ōuchi Yoshikuni. Town officials permitted photographs of the volume and of a picture map (p. 66-70).

In Miho, Endō Kunio kindly met with three North Americans to share with them "Miho yōji oboe" and how he came to possess it (p. 76). Mr. Endō's daughter, Mayumi, arranged a later gathering with two local historians, Endō Shōji and Watanabe Yasuhiro. She also provided copies of books on "Oboe" by Endō Shōji and others (p. 115).

Nagao Toshiyasu of Tokai University joined two of us in Miho for interviews of witnesses to the 1960 tsunami and 1974 storm: Shiba Tsune, Mizuno Teruko, and a lady in the Ishino family (p. 82-83). Moriguchi Osamu, of the central fire office of Shimizu city, arranged for an interview with another witness to the 1960 tsunami, Aoki Yukio.

Officials and residents of Tanabe welcomed us repeatedly for visits that included informative discussions with Kishi Akinori, a local historian, and field trips guided by

Shuto Nobuo at a memorial stone for the 1960 Chile tsunami near Miyako (map, p. 49). The inscription warns that even without an earthquake, a change in water level can mean a tsunami.

Moriai Miya of Tsugaruishi fields questions about flooding of her home by the 1960 Chile tsunami (p. 57, footnote). **Moriai Mutsuharu**, her neighbor, stands at right.

members of the city's general affairs office: its directors, Yamasaki Kiyohiro and Okamoto Yoshihiko, and staffers Urabe Shunji and Shin'ya Jun. Ōta Yūji, librarian with the municipal library, granted access to Tadokoro documents and shared his views of their history (p. 84-87). Hashimoto Kuniko and Minakata Fumie provided a tour of a Tanabe storehouse (photo, below left). In Shinjō, Matsuzaki Tomiji welcomed visitors to a storehouse site (p. 88) and Kashiwagi Tomio provided photos of the 1960 tsunami (p. 89).

Not far from Tanabe, in Hirogawa, Shimizu Isao gave three North Americans an enthusiastic, full-day field trip on Hamaguchi Goryō and his response to the 1854 tsunami that devastated Hiro-mura (photo, below right). At the time of that field trip, Mr. Shimizu was continuing education specialist at the town's community center, Hirogawa-chō Chūō Kōminkan. Tsumura Kenshiro, formerly of Hirogawa, further advised us on Goryō and "Inamura no hi." The picture on page 47 was taken by him and is reproduced with permission of the painting's owner, Yōgen Temple.

IN FORMER EDO, Watanabe Tokie of the Earthquake Research Institute (ERI), University of Tokyo, set up some of the rural visits. Murakami Yoshikane, while a graduate student at ERI, provided a speedy drive to northeast Japan. Katō Teruyuki of ERI advised us on tide-gauge data. Hirata Sakura and Kikuchi Ryōichi of Meiji University allowed us to examine maps of Japan and Suruga province from 1702 (p. 32, 76). Ota Yoko, formerly of Yokohama National University, helped us interpret the picture maps of Morioka-han (p. 36, 44, 58), the inland waterways between Nakaminato and Edo (p. 67), and land-level changes in northeast Japan (p. 65). She also arranged for an Edo mansion for Atwater and his family; and Joel Muraoka provided Tokyo lodging for Yamaguchi.

In nearby Tsukuba, Okada Masami and Tanioka Yūichirō of the Meteorological Research Institute, Japan Meteorological Agency, checked tidal measurements and datums. Odagiri Satoko, of the Geographical Survey Institute, provided old topographic maps. Staff of the Active Fault Research Center, a part of the National Institute of Advanced Industrial Science and Technology, extended countless courtesies to Atwater. These included telephone interviews and trip planning by Isoda Hisako, guidance on Japanese history and language from Horikawa Haruo and

Nanayama Futoshi, and bibliographic work by Satō Nobue. Dr. Horikawa photographed the monument on page 45; Ms. Satō, the anthologies on pages 62 and 123. Azuma Takashi led the visits to Hitachinaka and to the shogunal maps at Meiji University (p. 32, 76).

Atwater's contributions to the book were made possible, in part, by several visits to Japan. During the longest of these, for nearly a year, his travel and living expenses were covered by Japanese government fellowships from the Center for Global Partnership, ERI, the Science and Technology Agency, and the Geological Survey of Japan. Persons who made these fellowships possible include Usui Akira and Ozaki Hiromi of the Geological Survey of Japan; Satō Hiroshi, Shimazaki Kunihiko, and Murakami Tomoko of ERI; and Ruth Reid and Rebecca Barnhart, and Jack Medlin of the U.S. Geological Survey (USGS).

Matsuda Izumi welcomed Atwater to her first-year Japanese language course at the University of Washington. Yamaguchi drew on Japanese language training that includes a summer program in 1976 (sponsored by Sumitomo Bank) and immersion during an appointment at the Hokkaido Research Center of the Forestry and Forest Products Research Institute from 1994 to 1996 (supported by Japan's Science and Technology Agency).

THE NORTH AMERICAN PARENT for Japan's orphan tsunami of 1700 became known through the work of a great many people. The principals include Hiroo Kanamori of the California Institute of Technology; Tom Heaton and Alan Nelson of the USGS; and Minze Stuiver of the University of Washington.

The Nuclear Regulatory Commission underwrote the radiocarbon dating of trees and herbs killed by tidal submergence from the 1700 earthquake (p. 24-25). In Minze Stuiver's lab, Philip Wilkinson analyzed the spruce samples.

Unsung heroes of the earthquake's tree-ring dating include Boyd Benson, Lori Davis, John Shulene, Karl Wegmann, and Marco Cisternas, all of whom helped dig out and sample the stumps of earthquake-killed red cedar.

Pierre Saint-Amand provided sharp prints of the Chilean photos on pages 10 and 11. The Alaskan airphoto on page 14 comes from the collection of A. Thomas Ovenshine and Susan Bartsch-Winkler, formerly of the USGS. Ian Shennan supplied one of the more recent Alaskan images on page 95.

Hashimoto Kuniko leafs through a book from the collection of Minakata Kumagusu (1867-1941), a mycologist and folklorist. She stands in a traditional Tanabe storehouse on a floor 0.4-0.5 m above the ground. The 1700 tsunami may have flooded such a raised floor (**B**, p. 88).

Shimizu Isao of Hirogawa enumerates losses of life and property from the 1707 and 1854 tsunamis in Hiro village (p. 47).

THIS BOOK began in 1999 as a manuscript too large for its intially intended outlet, a volume of papers on subduction zones. Andō Masataka—who twenty years earlier published a seminal paper on Cascadia's great-earthquake potential—released Atwater from a promise to contribute to that volume.

Critical review began that year with Andrew Moore, then at Tohoku University, and Ruth Ludwin, University of Washington. Ebara Masaharu of the Historiographical Institute, University of Tokyo, corrected subsequent transliterations and translations of the Edo-period documents.

Later drafts were reviewed in full by Emile Okal of Northwestern University; Ruth Pelz of the Burke Museum, Seattle; Yoko Ota; and Ruth Kirk, Kip Ault, Eric Blackford, and an anonymous reader on behalf of University of Washington Press. Suggestions from the anonymous reader spurred reorganization of the book and expansion of its chapters on the Cascadia subduction zone. Additional reviewer were provided by Patricia Atwater, Lori Dengler, Adriana Erickson, Ned Field, Harumi Kato, Hayakawa Yukio, Hal Mojfeld, Joel Muraoka, Yoshiko Sorensen, and Vasily Titov. Pauline Curiel and Satō Nobue printed and circulated the reviewers' copies.

The book's cover was developed at University of Washington Press with design by Ashley Saleeba and editing by Alice Herzog and Marilyn Trueblood. Sophia Smith and Pat Soden contributed to the book's English title.

The reference list includes titles located by Keiko Yokota-Carter, the Japanese-language specialist at the East Asia Library, University of Washington. Inoue Megumi, Nakamura Noriyuki, and Ekida Fusae, bilingual graduate students from Japan, translated reference materials and romanized bibliographic citations. Additional translations were provided by Tajima Maiko and Harada Shino. Annaliese Eipert helped compile the references.

The book's design is based on a USGS pamphlet by Peter Ward, Robert Page, Laurie Hodgen, and Jeff Troll, and on examples presented by Edward Tufte. Susan Mayfield and Sara Boore of the USGS provided guidance on color, fonts, and layout; Boore also prepared the block diagrams adapted on page 10. Ed Mulligan and Lorien Freeman, University of Washington, helped us mock up pages by providing computer-network connections and maintaining a color printer.

The USGS granted Atwater freedom to devote several years to the book. Michael Blanpied, Nancy Rountree, Peter Stauffer, and Jane Ciener helped set aside USGS funds for editing and printing. Ruth Kirk initiated discussions, with Michael Duckworth and Pat Soden, that led to joint publication by University of Washington Press.

PERMISSIONS to reproduce proprietary maps, prints, paintings, and photographs were granted by the following institutions:

The Art Institute of Chicago—p. 71
Asahi Shimbun—p. 55
Gakken Co.—p. 115, 117, 119, 121
Hiro Elementary School—p. 113
Hitachinaka city—p. 66, 131
Meiji University Library, Tokyo—p. 32-33, 76
Morioka City Central Community Center, Documents Office—p. 36, 44, 45, 49, 50, 58
Ōfunato city—p. 81, 133
Sendai Museum—p. 127
Shinjō Community Center—p. 85, 89
Tanabe Municipal Library—p. 84, 90
United States National Archives and Records Administration—p. 12
University of California, Berkeley, East Asian Library—p. 26, 30-31, 41, 43, 61, 70-72, 76, 106, back cover
University of California, Berkeley, National Information Service for Earthquake Engineering—p. 9
University of Washington Libraries, Special Collections—frontispiece and p. 2, 13, 104, 129
Yōgen temple, Hirogawa—p. 47

The paper by Andō Masataka, with Emery Balazs, initiated the study of cyclic land-level change at Cascadia (Ando and Balazs, 1979). We consulted Ward and others (1989) and Tufte (1990, 2001) on book design.

The Nuclear Regulatory Commission, reviewing the design of this power plant, supported carbon-14 dating of Cascadia earthquakes (p. 25).

Satsop, Washington (location map, p. 96).

Boyd Benson, in an Oregon tidal swamp, checks the annual rings of a spruce survivor of the 1700 earthquake (p. 97).

Authors 著者紹介

Brian Atwater, Musumi-Rokkaku Satoko, Satake Kenji, Tsuji Yoshinobu, Ueda Kazue, and David Yamaguchi. Tokyo, 2004.

THE STORY OF THE 1700 TSUNAMI draws on human history interpreted from old Japanese documents, on natural history inferred from North American sediments, trees, and native legends, and on mathematical modeling of tsunamis. The authors pooled their backgrounds in these and other fields. Below, as on the cover and title page, their names appear alphabetically.

Brian F. ATWATER ブライアン・F・アトウォーター conceived of the book and led in its preparation. To this work he brought over a decade of experience with geologic records of the 1700 earthquake and tsunami in North America. Through field work in 1999 he also contributed to size estimates for the 1700 tsunami in Japan. He holds B.S. and M.S. degrees in geology from Stanford University and a Ph.D. in geology from the University of Delaware. In thirty years with the U.S. Geological Survey he has studied bay and river geology in California, ice-age floods in Washington, and geologic records of earthquakes and tsunamis in the United States, Chile, and Japan. He lives in Seattle and is based at the University of Washington.

MUSUMI-ROKKAKU Satoko 六角 聰子 guided the transliteration and translation of the tsunami accounts. She also contributed to interviews in northeast Japan and to historical background material. Her education includes a B.A. in Humanities at Tokyo's International Christian University and an ensuing year as a Fulbright Fellow at the University of Chicago, where she did graduate work in Islamic cultural history and Arabic language. Since 1979 she has coordinated the United Nations University fellowship program for Asian food scientists while teaching at Tokyo's Obirin University. She has served as an officer in the UNU Women's Association and holds an honorary professorship at the Mongolian University of Science and Technology. Her travels have taken her to 33 countries.

SATAKE Kenji 佐竹 健治 estimated sizes of the 1700 tsunami in Japan and the 1700 earthquake at Cascadia. He also tracked down primary sources for accounts of the 1700 tsunami in Tsugaruishi and Nakaminato. These contributions stem from his broad interest in subduction-zone earthquakes, which he studies with instrumental, written, and geological records, and with geophysical modeling. He holds B.S. and M.S. degrees in geophysics from Hokkaido University and a Ph.D. in geophysics from the University of Tokyo. He spent seven years in the United States, as a postdoctoral researcher at the California Institute of Technology and as an assistant professor at the University of Michigan. Since 1995 he has worked at the Geological Survey of Japan, where he is now deputy director of the Active Fault Research Center of the National Institute of Advanced Industrial Science and Technology. His field work in 2005 included post-tsunami surveys in Myanmar and Thailand. He chairs the tsunami commission of the International Union of Geodesy and Geophysics, serves on governmental committees that evaluate earthquake hazards in Japan, and edits "Rekishi Jishin," the journal of Japan's Society of Historical Earthquake Studies.

TSUJI Yoshinobu 都司 嘉宣 identified places reached by the 1700 tsunami, computed tides for estimates of the tsunami's height, and helped transliterate and translate the tsunami accounts. From the University of Tokyo he earned a B.S. in civil engineering, and M.S. and Ph.D. degrees in geophysics. His studies of Japan's historical earthquakes and tsunamis began in the 1970s, when he worked for the National Research Center for Disaster Prevention. In 1987 he joined the faculty of the University of Tokyo's Earthquake Research Institute. He subsequently participated in post-tsunami field surveys in Nicaragua and Papua New Guinea, and he led such surveys in 2005 in Aceh and Thailand. He has also investigated storm surges and tsunami-induced damage to buildings. His second languages include Korean, Chinese, Russian, English, and Fortran.

UEDA Kazue 上田 和枝 discovered, transliterated, and translated accounts of the 1700 tsunami. She also confirmed the tsunami's misdating in Moriai-ke "Nikki Kakitome-chō" (p. 53), investigated the historical context of the tsunami's accounts, and interviewed witnesses to the 1960 tsunami. For over thirty years she has specialized in the written records of Japanese earthquakes. She entered that field eleven years after earning a B.A. in psychology at Tokyo Woman's Christian College and joining the Earthquake Research Institute, University of Tokyo. The 21-volume, 16,812-page earthquake anthology, "Shinshū Nihon jishin shiryō" (p. 123), resulted largely from her efforts. These included some 300 visits to libraries, prominent families, government offices, temples, and shrines where she searched thousands of pages daily for accounts of earthquakes and tsunamis. Since retiring from the Earthquake Research Institute in 1998 she has remained active in meetings and publications on Japan's historical earthquakes.

David K. YAMAGUCHI デイビッド・K・ヤマグチ relentlessly revised the entire book for presentation and content. He also contributed tree-ring dates, photographs, and bilingual interviews in Tsugaruishi, Miho, and Tanabe. A Seattle-born grandson of Japanese immigrants, he earned a B.S. in biology at Yale and a Ph.D. in forestry at the University of Washington. While a graduate student, he dated two eruptions of Mount St. Helens to 1479-1482 from the thin rings of trees damaged downwind. These findings led to a postdoctoral fellowship with the U.S. Geological Survey, where he dated volcanic debris flows by matching the ring-width patterns of entombed trees with those of living ones. During that fellowship he began the coastal tree-ring studies that helped identify Cascadia as the source of the orphan tsunami (p. 24, 96-97). Those studies progressed while he served on the research faculty of the University of Colorado and worked as a visiting scholar at the Forestry and Forest Products Research Institute, Hokkaido. Later he became a financial advisor at Merrill Lynch and a public-health statistician at the University of Washington's School of Dentistry. He now analyzes public-health data as a programmer at the Center for Health Studies, Group Health Cooperative, Seattle.

PLEASE SEND CORRECTIONS to the corresponding author or authors identified at http://pubs.usgs.gov/pp/pp1707/.

＊〔御番所日記〕
十月十三日　午ノ下刻少地震、主膳同公、

元禄十二年十月二十四日（西暦一六九九、一二、一四、）日光地少シク震フ、
＊〔御番所日記〕
十月廿四日　未下刻少地震、

元禄十二年十一月三日（西暦一六九九、一二、二三、）日光地少々震フ、
＊〔御番所日記〕
十一月三日　寅ノ后刻少地震

元禄十二年十一月十七日（西暦一七〇〇、）日光地少々震フ、
＊〔御番所日記〕
十一月十七日　辰ノ中刻少地震、御宮中無御別條、

元禄十二年十一月十九日（西暦一七〇〇、）日光地震強シ、
＊〔御番所日記〕
十一月十九日　丑后刻地震餘程、御宮中巡見仕候処二御別條無之、仁王御門又当番衆不残大衆院皆々出仕、御目付分御使來ル、其外役人衆伺公、

元禄十二年十二月一日（西暦一七〇〇、）日光地震フ、
＊〔御番所日記〕
極月朔日　未后刻地震　略　下

＊〔御番所日記〕
極月朔日　未后刻地震　略　下

元禄十二年十二月八日（西暦一七〇〇、一二、二七、）紀伊國潮汐常ニ異ナレリ、
＊〔田辺町大帳〕○紀
十二月八日潮水非常ニ増長

元禄十二年十二月八日（西暦一七〇〇、一二、二七、）十二月陸中大槌津浪アリ、
＊〔大槌記録抄〕
この年極月夜九ツに大汐きし、海辺大分驚きす、人馬怪我なし

元禄十二年十二月廿三日（西暦一七〇〇、一二、三〇、）加賀國金沢卯辰山再ビ崩壞シ、壓死者三十一人・家屋倒壞若干ヲ出セリ、
＊〔前田家東〕十二月廿三日申刻、卯辰山破壞シ、浅野川ヲ塞グ、壓死三十一人、
＊〔三州志來因觀覽附録〕十二月二十三日申時、城東茶臼山崩壞シ、浅野川塞ル、壓死三十一人、壓家若干也、
＊〔愛異記〕十二月廿三日申ノ刻、卯辰茶臼山崩テ浅野川中、町夫毎日千人ヲ以テ晝夜水道ヲ鑿ツ、

（二五）

Two accounts of the 1700 tsunami were first published in Musha Kinkichi's second volume of collected materials on Japanese historical earthquakes. The accounts, boxed above, are quoted on page 62. The volume is listed in the references by its corporate author, Mombushō Shinsai Yobō Hyōgikai.

Mombushō Shinsai Yobō Hyōgikai (1943, p. 25).

A

Abe, K., 1979, Size of great earthquakes of 1837-1974 inferred from tsunami data: Journal of Geophysical Research, v. 84, p. 1561-1568. {*footnoted on our page* 49}

Adams, J., 1990, Paleoseismicity of the Cascadia subduction zone—evidence from turbidites off the Oregon-Washington margin: Tectonics, v. 9, p. 569-583. {22, 101}

Adams, J., and Atkinson, G., 2003, Development of seismic hazard maps for the proposed 2005 edition of the National Building Code of Canada: Canadian Journal of Civil Engineering, v. 30, p. 255-271. {105}

Akioka, T., 1997, Nihon chizu shi [History of Japanese maps]: Tokyo, Myūjiamu Tosho, 339 p. [in Japanese] {31}

All Japan Handmade Washi Association, 1991, Handbook on the art of washi: Tokyo, Wagami-do K.K., 125 p. {87}

Ando, M., 1975, Source mechanisms and tectonic significance of historical earthquakes along the Nankai Trough, Japan: Tectonophysics, v. 27, p. 119-140. {85, 91, 101}

Ando, M., and Balazs, E.I., 1979, Geodetic evidence for aseismic subduction of the Juan de Fuca plate: Journal of Geophysical Research, v. 84, p. 3023-3028. {109}

Andō, S., and Wakayama-ken Tanabe-shi Kyōiku I'inkai, editors, 1991-1994, Kishū Tanabe mandaiki: Osaka, Seibundō Shuppan, 18 volumes, 10,200 p. [in Japanese]. {84}

Anonymous, editor, 1995, Sumpuki: Tokyo, Zoku Gunshoruijū Kanseikai, 319 p. [in Japanese]. {41}

Arakawa, H., and Taga, S., 1969, Climate of Japan, *in* Arakawa, H., Climates of northern and eastern Asia: Amsterdam, Elsevier, World survey of climatology v. 8, p. 119-158. {72, 83}

Arakawa, H., Ishida, Y., and Itō, T., editors, 1961, Nihon takashio shiryō [Historical materials on storm surges in Japan]: Meteorological Research Institute, 272 p. [in Japanese]. {83}

Ashida Bunko Hensan I'inkai [Ashida Collection Editorial Committee], 2004, Ashida bunko mokuroku kochizu hen [Index to the Ashida collection of old maps]: Tokyo, Meiji University, Jimbunkagaku Kenkyūsho [Humanities Research Center], 329 p. [in Japanese]. {32, 76}

Aston, W.G., translator, 1972, Nihongi; chronicles of Japan from the earliest times to A.D. 697, volume 2: Rutland, Vermont, and Tokyo, Charles E. Tuttle, 443 p. {54}

Atkinson, G.M., and Boore, D.M, 2003, Empirical ground-motion relations for subduction-zone earthquakes and their application to Cascadia and other regions: Bulletin of the Seismological Society of America, v. 93, p. 1703-1729. {104}

Atkinson, G.M., and Casey, R., 2003, A comparison of ground motions from the 2001 M 6.8 in-slab earthquakes in Cascadia and Japan: Bulletin of the Seismological Society of America, v. 93, p. 1823-1831. {104}

Atwater, B.F., and Hemphill-Haley, E., 1997, Recurrence intervals for great earthquakes of the past 3,500 years at northeastern Willapa Bay, Washington: U.S. Geological Survey Professional Paper 1576, 108 p. {18, 21, 24}

Atwater, B.F., and Yamaguchi, D.K., 1991, Sudden, probably coseismic submergence of Holocene trees and grass in coastal Washington State: Geology, v. 19, p. 706-709. {17}

Atwater, B.F., Stuiver, M., and Yamaguchi, D.K., 1991, Radiocarbon test of earthquake magnitude at the Cascadia subduction zone: Nature, v. 353, p. 156-158. {25}

Atwater, B.F., Yelin, T.S., Weaver, C.S., and Hendley, J.W., III, 1995, Averting surprises in the Pacific Northwest: U.S. Geological Survey Fact Sheet 111-95, 2 p. [http://quake.wr.usgs.gov/prepare/factsheets/PacNW/]. {104}

Atwater, B.F., Cisternas V., M., Bourgeois, J., Dudley, W.C., Hendley, J.W. II, and Stauffer, P.H., 1999, Surviving a tsunami—lessons from Chile, Hawaii, and Japan: U.S. Geological Survey Circular 1187, 18 p. [pubs.usgs.gov/circ/c1187/; in Spanish, as U.S. Geological Survey Circular 1218, http://pubs.usgs.gov/circ/c1218/]. {5, 11, 49, 80}

Atwater, B.F., Baker, D., Barnhardt, W.A., Burrell, K.S., Haraguchi, T., Higman, B., Kayen, R.E., Minasian, D., Nakata, T., Satake, K., Shimokawa, K., Takada, K., and Cisternas V., M., 2001a, Grouted sediment slices show signs of earthquake shaking: Eos, v. 82, p. 603, 608. {23}

Atwater, B.F., Yamaguchi, D.K., Bondevik, S., Barnhardt, W.A., Amidon, L.J., Benson, B.E., Skjerdal, G., Shulene, J.A., and Nanayama, F., 2001b, Rapid resetting of an estuarine recorder of the 1964 Alaska earthquake: Geological Society of America Bulletin, v. 113, p. 1193-1204. {14}

Atwater, B.F., Tuttle, M.P., Schweig, E.S., Rubin, C.M., Yamaguchi, D.K., and Hemphill-Haley, E., 2004, Earthquake recurrence inferred from paleoseismology, *in* Gillespie, A.R., Porter, S.C., and Atwater, B.F., editors, The Quaternary Period in the United States: Amsterdam, Elsevier, p. 331-350. {16, 100, 101}

B

Bache, A.D., 1856, Notice of earthquake waves on the western coast of the United States, on the 23d and 25th of December, 1854: American Journal of Science and Arts, second series, v. 21, p. 37-43. [text published also in Report of the Superintendent of the Coast Survey showing the progress of the Survey during the year 1855: 34th Congress, 1st Session, Ex. Doc. 22, p. 342-346 http://docs.lib.noaa.gov/rescue/cgs/001_pdf/CSC-0004.pdf] {91}

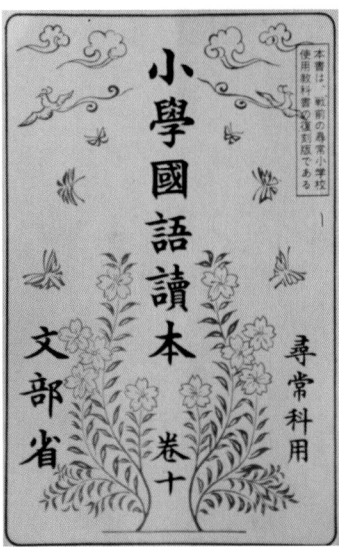

A village elder saves his fellow villagers from a tsunami in a story that was introduced to Japanese school-children in the 1930s and 1940s (textbook cover, left). Inspired by real events in 1854 (p. 47), the plot runs through video frames on succeeding pages. Related references are listed under the author names Hearn, Hodges, Shimizu, and Tsumura.

CITATION ORDER for entries with the same first author (Jones): (1) publications by Jones alone, arranged by year; (2) publications by Jones and one other author, arranged alphabetically by second author, then by year; and (3) papers by Jones and two or more coauthors, arranged by year alone. In the book's footnotes, these publications would be attributed to (1) Jones, (2) Jones and Smith, and (3) Jones and others.

DIVISION BETWEEN TITLE AND SUBTITLE is marked here by a semi-colon if denoted in the original by a colon or font change.

Ballantyne, D., Bartoletti, S., Chang, S., Graff, B., MacRae, G., Meszaros, J., Pearce, I., Pierepiekarz, M., Preuss, J., Stewart, M., Swanson, D., and Weaver, C., 2005, Scenario for a magnitude 6.7 earthquake on the Seattle fault: Earthquake Engineering Research Institute and Washington Emergency Management Division [http://seattlescenario.eeri.org/documents/EQ%202-28%20Booklet.pdf]. {104}

Bartsch-Winkler, S.R., 1988, Cycle of earthquake-induced aggradation and related tidal channel shifting, upper Turnagain Arm, Alaska, USA: Sedimentology, v. 35, p. 621-628. {14}

Beasley, W.G., 1982, The modern history of Japan, third revised edition: Tokyo, Charles E. Tuttle, 358 p. {45}

Beck, J.L., and Hall, J.F., 1986, Factors contributing to the catastrophe in Mexico City during the earthquake of September 19, 1985: Geophysical Research Letters, v. 13, p. 593-596. {8}

Ben-Menahem, A., and Rosenman, M., 1972, Amplitude patterns of tsunami waves from submarine earthquakes: Journal of Geophysical Research, v. 77, p. 3097-3128. {54}

Benson, B.E., Atwater, B.F., Yamaguchi, D.K., Amidon, L.J., Brown, S.L., and Lewis, R.C., 2001, Renewal of tidal forests in Washington state after a subduction earthquake in A.D. 1700: Quaternary Research, v. 56, p. 139-147. {96}

Bernard, E.N, Mader, C., Curtis, G., and Satake, K., 1994, Tsunami inundation model study of Eureka and Crescent City, California: National Oceanic and Atmospheric Administration, Environmental Research Laboratories, Pacific Marine Environmental Laboratory, PMEL Technical Memorandum ERL PMEL 103, 80 p. {102}

Berry, M.E., 1982, Hideyoshi: Cambridge, Mass., Harvard University Press, 293 p. {86}

Bilham, R., Engdahl, R., Feldl, N., and Satyabala, S.P., 2005, Partial and complete rupture of the Indo-Andaman plate boundary 1847-2004: Seismological Research Letters, v. 76, p. 299-311. {5, 101}

Bolitho, H., 1976, The dog shogun, in Wang, G.W., editor, Self and biography, essays on the individual and society in Asia: Sydney, Sydney University Press for the Australian Academy of the Humanities, p. 123-139. {63}

Bolitho, H., 1991, The han, in Hall, J.W., editor, and McClain, J.L., assistant editor, The Cambridge history of Japan (Hall, J.W., Jansen, M.B., Kanai, M., and Twitchett, D., general editors), volume 4, early modern Japan: Cambridge, U.K., Cambridge University Press, p. 183-234. {61}

Boudonnat, L., and Kushizaki, H., 2003, Traces of the brush: the art of Japanese calligraphy: Paris, Éditions du Seuil, and San Francisco, Chronicle Books, 215 p. [translated by C. Penwarden from Au fil du pinceau, la calligraphie japonaise]. {87}

Brocher, T.M., Parsons, T., Tréhu, A.M., Snelson, C.M., and Fisher, M.A., 2003, Seismic evidence for widespread serpentinized forearc upper mantle along the Cascadia margin: Geology, v. 31, p. 267-270. {104}

Bryant, E., 2001, Tsunami; the underrated hazard: Cambridge, U.K., Cambridge University Press, 320 p. {11}

Bucknam, R.C., Hemphill-Haley, E., and Leopold, E.B., 1992, Abrupt uplift within the past 1700 years at southern Puget Sound, Washington: Science, v. 258, p. 1611-1614. {104}

Building Seismic Safety Council, 2001, NEHRP recommended provisions (National Earthquake Hazard Reduction Program) for seismic regulations for new buildings and other structures, 2000 edition, part 1: provisions (FEMA [Federal Emergency Management Agency] 368): Washington, D.C., Building Seismic Safety Council, 374 p. {104}

C

Cascadia Region Earthquake Workgroup, 2005, Cascadia subduction zone earthquakes; a magnitude 9.0 earthquake scenario: Seattle, Cascadia Region Earthquake Workgroup, 21 p. [http://www.crew.org/papers/CREWCascadiaFinal.pdf] [also available as Oregon Department of Geology and Mineral Industries Open-file Report 05-05]. {5, 104}

Cazenave, A., and Nerem, R.S., 2004, Present-day sea level change; observations and causes: Reviews of Geophysics, v. 42, 2003RG000139. {65}

The Central Meteorological Observatory, 1953, Shōwa 27-nen 11-gatsu Kamchatka jishin chōsa hōkoku [Report of investigation of the 1952 Kamchatka earthquake]: Quarterly Journal of Seismology, v. 18, no. 1, 48 p. [in Japanese]. {37, 51, 54, 59, 94, 95}

Chamberlain, B.H., 1905, Things Japanese; being notes on various subjects connected with Japan for the use of travellers and others, fifth edition revised: London, J. Murray, 568 p. [reprinted, as "Japanese things," in 1971 by Charles E. Tuttle, Co., Rutland, Vermont, and Tokyo]. {43, 87}

Chibbett, D.G., 1977, The history of Japanese printing and book illustration: Tokyo and New York, Kodansha, 264 p. {29}

Cifuentes, I., 1989, The 1960 Chilean earthquakes: Journal of Geophysical Research, v. 94, p. 665-680. {9}

Cisternas [V.], M., Atwater, B.F., Torrejón, F., Sawai, Y., Machuca, G., Lagos, M., Eipert, A., Youlton, C., Salgado, I., Kamataki, T., Shishikura, M., Rajendran, C.P., Malik, J.K., Rizal, Y., and Husni, M., 2005, Predecessors to the giant 1960 Chile earthquake: Nature, v. 437, p. 404-407. {19}

Clague, J.J., and Bobrowsky, P.T., 1994, Tsunami deposits beneath tidal marshes on Vancouver Island, British Columbia: Geological Society of America Bulletin, v. 106, p. 1293-1303. {101}

Clague, J.J., Bobrowsky, P.T., and Hutchinson, I., 2000, A review of geological records of large tsunamis at Vancouver Island, British Columbia, and implications for hazard: Quaternary Science Reviews, v. 19, p. 849-863. {18}

Clarke, S.H., Jr., and Carver, G.A., 1992, Late Holocene tectonics and paleoseismicity, southern Cascadia subduction zone: Science, v. 255, p. 188-192. {101}

Coastal Movements Data Center, 1996, Tables and graphs of annual mean sea level along the Japanese coast 1894~1995: Tsukuba, Geographical Survey Institute, 113 p. {65}

Cole, S.C., Atwater, B.F., McCutcheon, P.T., Stein, J.K., and Hemphill-Haley, E., 1996, Earthquake-induced burial of archaeological sites along the southern Washington coast about A.D. 1700: Geoarchaeology, v. 11, p. 165-177. {20}

Combellick, R.A., 1991, Paleoseismicity of the Cook Inlet region, Alaska; evidence from peat stratigraphy in Turnagain and Knick Arms: Alaska Division of Geology and Geophysical Surveys Professional Report 112, 52 p. {95}

The Committee for Field Investigation of the Chilean Tsunami of 1960, 1961, Report on the Chilean tsunami of May 24, 1960, as observed along the coast of Japan: Tokyo, Maruzen Co., Ltd., 1961, 397 p. {37, 49, 51, 55, 56, 57, 73, 83}

Cooper, J.G., 1860, Report upon the botany of the route, in Reports of explorations and surveys to ascertain the most practicable and economical route for a railroad from the Mississippi River to the

Pacific Ocean, v. XII, book II: 36th Congress, 1st session, House of Representatives, Ex. doc. no. 56, p. 13-39. {16}

Cox, D.C., 2001, The inappropriate tsunami icon: Science of Tsunami Hazards, v. 19, p. 87-92 [http://epubs.lanl.gov/tsunami/5092.pdf]. {80}

D

Darienzo, M.E., and Peterson, C.D., 1995, Magnitude and frequency of subduction-zone earthquakes along the Oregon coast in the past 3,000 years: Oregon Geology, v. 57, p. 3-12. {101}

Doig, I., 1980, Winter brothers; a season at the edge of America: New York, Harcourt Brace Jovanovich, 246 p. {12}

Dragert, H., and Hyndman, R.D., 1995, Continuous GPS monitoring of elastic strain in the northern Cascadia subduction zone: Geophysical Research Letters, v. 22, p. 755-758. {99}

Dragert, H., Wang, K., and James, T.S., 2001, A silent slip event on the deeper Cascadia subduction interface: Science, v. 292, p. 1525-2528. {99}

Dudley, W.C., and Lee, M., 1998, Tsunami! Honolulu, University of Hawai'i Press, 362 p. {11}

E

Earthquake Research Committee, 1998, Seismic activity in Japan — regional perspectives on the characteristics of destructive earthquakes — (excerpt): Tokyo, Science and Technology Agency, 222 p. (prepared by Earthquake Research Committee, Headquarters for Earthquake Research Promotion, Prime Minister's Office; translated by Earthquake Research Center, Association for the Development of Earthquake Prediction). {65}

Endō, S., and Nagasawa, K., editors, 1989, Miho chiku no rekishi, soko ga shiritai [Miho district history, what I want to know]: Shimizu-shi Miho Kōminkan [Shimizu City, Miho Community Center] and Miho Chiku Machi Zukuri Suishin I'inkai [Miho District Town Development Promotion Commission], 48 p. + appendices [in Japanese]. {76, 82}

Endō, S., Nagasawa, K., and Suzuki, M., editors, 1990, Miho chiku no rekishi, soko ga shiritai [Miho district history, what I want to know]: Shimizu-shi Miho Kōminkan [Shimizu City, Miho Community Center] and Miho Chiku Machi Zukuri Suishin I'inkai [Miho District Town Development Promotion Commission], 121 p. + appendices [in Japanese]. {76}

F

Frankel, A.[D.], Mueller, C., Barnhard, T., Perkins, D., Leyendecker, E.V., Dickman, N., Hanson, S., and Hopper, M., 1996, National seismic hazard maps, June 1996 documentation: U.S. Geological Survey Open-File Report 96-532, 69 p. [http://eqhazmaps.usgs.gov/hazmapsdoc/junecover.html] {105}

Frankel, A.D., Carver, D.L., and Williams, R.A., 2002a, Nonlinear and linear site response and basin effects in Seattle for the M 6.8 Nisqually, Washington, earthquake: Bulletin of the Seismological Society of America, v. 92, p. 2090-2109. {104}

Frankel, A.D., Petersen, M.D., Mueller, C.S., Haller, K.M., Wheeler, R.L., Leyendecker, E.V., Wesson, R.L., Harmsen, S.C., Cramer, C.H., Perkins, D.M., and Rukstales, K.S., 2002b, Documentation for the 2002 update of the national seismic hazard maps: U.S. Geological Survey Open-File Report 02-420, 33 p. [http://pubs.usgs.gov/of/2002/ofr-02-420/] {105}

French, J., editor, 1999, Tooley's dictionary of mapmakers, revised edition A-D: Herts, England, Map Collector Publications, 408 p. {5}

Fritts, H.C., 1976, Tree rings and climate: London, Academic Press, 576 p. [reprinted in 2001 by Blackburn Press, Caldwell, New Jersey]. {97}

G

García, V., and Suárez, G., 1996, Los sismos en la historia de México: México [City], Universidad Nacional Autónoma de México, 718 p. [in Spanish]. {94}

Garruth, G., 1993, The encyclopedia of world facts and dates: New York, Harper Collins, 1310 p. {5}

Goldfinger, C., Nelson, C.H., and Johnson, J.E., 2003, Holocene earthquake records from the Cascadia subduction zone and northern San Andreas fault based on precise dating of offshore turbidites: Annual Reviews of Earth and Planetary Science, v. 31, p. 555-577. {22}

González, F.I., 1984, Case study of wave-current-bathymetry interactions at the Columbia River entrance: Journal of Physical Oceanography, v. 14, p. 1065-1078. {73}

Goodman, L.J., and Swan, H., 2003, Singing the songs of my ancestors; the life and music of Helma Swan, Makah elder: Norman, Oklahoma, University of Oklahoma Press, 339 p. {12}

Grant, W.C., 1992, Paleoseismic evidence for late Holocene episodic subsidence on the northern Oregon coast: Seattle, University of Washington unpublished M.S. non-thesis report, 39 p. {21}

H

Hachinohe Komonjo Benkyōkai [Hachinohe Old-Documents Study Group], editor and publisher, 1994, Nambu-han "Han nikki" metsukesho kanjōsho kaidoku [Deciphering the official records of the inspection bureau and finance office of the Nambu Hachinohe domain], *in* Shoshanbun [Transcribed documents], v. 5 [for the year Genroku 12]: Hachinohe, Aomori Prefecture, 208 p. {52}

Hall, J.W., 1991, Introduction, *in* Hall, J.W., editor, and McClain, J.L., assistant editor, The Cambridge history of Japan (Hall, J.W., Jansen, M.B., Kanai, M., and Twitchett, D., general editors), volume 4, early modern Japan: Cambridge, U.K., Cambridge University Press, p. 1-39. {86}

Hall, R., and Radosevich, S.C., 1998, Geoarchaeological analysis of a site in the Cascadia subduction zone on the southern Oregon coast: Northwest Anthropological Research Notes, v. 29, p. 123-140. {20}

Hamilton, S.L., and Shennan, I., 2005, Late Holocene land and sea-level changes and the earthquake deformation cycle around upper Cook Inlet, Alaska: Quaternary Science Reviews, v. 24, doi:10.1016/j.quascirev.2004.11.003. {95}

One autumn evening at his clifftop home, a village headman feels an earthquake. Fearing a tsunami, he studies the sea.

Hanasaka, K., editor, 1974, Miyako no ayumi [Historical development of Miyako]: Miyako Kyōdoshi Henshū I'inkai [Miyako Document Editing Committee], Miyako Shiyakusho [Miyako City Hall], 126 p. [in Japanese]. {39, 49}

Hanley, S.B., and Yamamura, K., 1977, Economic and demographic change in preindustrial Japan, 1600-1868: Princeton, N.J., Princeton University Press, 409 p. {36, 44, 53, 61}

Haring, C.H., 1963. The Spanish empire in America: New York, Harcourt, 371 p. {94}

Harmsen, S.C., Frankel, A.D., and Petersen, M.D., 2003, Deaggregation of U.S. seismic hazard sources: the 2002 update: U.S. Geological Survey Open-file Report 03-440, 33 p. [http://pubs.usgs.gov/of/2003/ofr-03-440/]. {105}

Hatori, T., 1965, On the Alaska tsunami of March 28, 1964, as observed along the coast of Japan: Bulletin of the Earthquake Research Institute, v. 43, p. 399-408. {54, 94, 95}

Hatori, T., 1976, Documents of tsunami and crustal deformation in Tokai district associated with the Ansei earthquake of Dec. 23, 1854: Bulletin of the Earthquake Research Institute, v. 51, p. 13-28 [in Japanese with English abstract and captions]. {77, 82}

Hatori, T., 1995, Field investigation of the 1611 Keicho [Keichō] Sanriku tsunami along the Iwate coast, NE Japan: Rekishi Jishin [Historical Earthquakes], v. 11, p. 59-66 [in Japanese with English captions]. {37, 41, 51, 59}

Hayes, D., 1999, Historical atlas of the Pacific Northwest; maps of exploration and discovery; British Columbia, Washington, Oregon, Alaska, Yukon: Seattle, Sasquatch Books, 208 p. [reprinted with revision, 2000]. {12}

Hearn, L., 1897, Gleanings in Buddha-fields: studies of hand and soul in the Far East: Boston, Houghton, Mifflin, 296 p. [reprinted in 1971 by Charles E. Tuttle, Co., Rutland, Vermont, and Tokyo, 296 p.] {47}

Heaton, T.H., and Hartzell, S.H., 1986, Source characteristics of hypothetical subduction earthquakes in the northwestern United States: Bulletin of the Seismological Society of America, v. 76, p. 675-708. {8}

Heaton, T.H., and Hartzell, S.H., 1987, Earthquake hazards on the Cascadia subduction zone: Science, v. 236, p. 162-168. {8}

Heaton, T.H., and Hartzell, S.H., 1989, Estimation of strong ground motions from hypothetical earthquakes on the Cascadia subduction zone, Pacific Northwest: Pure and Applied Geophysics, v. 129, p. 131-201. {104}

Heki, K., 2004, Space geodetic observation of deep basal subduction erosion in northeastern Japan: Earth and Planetary Science Letters, v. 219, p. 13-20. {65}

Hemphill-Haley, E., 1996, Diatoms as an aid in identifying late-Holocene tsunami deposits: The Holocene, v. 6, p. 439-448. {18}

Hibiya, T., and Kajiura, K., 1982, Origin of the abiki phenomenon (a kind of seiche) in Nagasaki Bay: Journal of the Oceanographical Society of Japan, v. 38, p. 172-182. [http://www.terrapub.co.jp/journals/JO/JOSJ/toc/3803.html]. {86}

Hodges, M., 1964, The wave; adapted from Lafcadio Hearn's Gleanings in Buddha-fields, illustrated by Blair Lent: Boston, Houghton Mifflin Co., 45 p. [reprinted in 1997 by Harcourt School Publishers, Orlando, Florida, 48 p.]. {47}

Hoshikawa, M., and Maezawa, T., 1984-1985, Nambu-han sankō sho-kakei-zu [Family histories in Nambu-han]: Tokyo, Kokusho Kankō-kai, five volumes, 2,809 p. [in Japanese]. {44}

Hosoi, K., 1988, Morioka-han, in Kimura, M., Fujino, T., and Murakami, T., editors, Han-shi daijiten dai 1-kan, Hokkaido,

Tōhoku hen [Encyclopedia of the history of ruling clans and their Edo-period domains, v. 1, Hokkaido and Tōhoku]: Tokyo, Yūzan Kaku, p. 56-80 [in Japanese]. {44, 61}

Hutchinson, I., Clague, J.J., and Mathewes, R.W., 1997, Reconstructing the tsunami record on an emerging coast; a case study of Kanim Lake, Vancouver Island, British Columbia: Journal of Coastal Research, v. 13, p. 545-553. {18}

Hyndman, R.D., 1995, Giant earthquakes of the Pacific Northwest: Scientific American, v. 273, p. 50-57. {8}

I

Ishibashi, K., 1981, Specification of a soon-to-occur seismic faulting in the Tokai district, central Japan, based upon seismotectonics, in Simpson, D.W., and Richards, P.G., editors, Earthquake prediction, an international review: American Geophysical Union, Maurice Ewing series 4, p. 297-332. {77, 85}

Ishibashi, K., 1984, Coseismic vertical crustal movements in the Suruga Bay region: Daiyonki Kenkyū (The Quaternary Research), v. 23, p. 105-110. [in Japanese with English abstract]. {82}

Iwamoto, Y., 1970, Kinsei gyoson kyōdōtai no hensen katei — shōhin keizai no shinten to sonraku kyōdōtai — [Changing course of early modern fishing village communities— development of commercial economy and village communities]: Tokyo, Hanawa Shobō, 256 p. [in Japanese]. {36, 51, 53, 57}

Iwamoto, Y., 1979, Nambu hanamagari no sake [Nambu crook-nose salmon]: Tokyo, Nihon Keizai Shimbunsha, 245 p. [in Japanese]. {53}

J

Jacoby, G.[C.], Carver, G.[A.], and Wagner, W., 1995, Trees and herbs killed by an earthquake ~300 yr ago at Humboldt Bay, California: Geology, v. 23, p. 77-80. {17}

Jacoby, G.C., Bunker, D.E., and Benson, B.E., 1997, Tree-ring evidence for an A.D. 1700 Cascadia earthquake in Washington and northern Oregon: Geology, v. 25, p. 999-1002. {96, 97}

Japan Meteorological Agency, 1960, Tide tables for the year 1961: Japan Meteorological Agency, 402 p. [in Japanese and English]. {48}

Japan Meteorological Agency, 1961, The report on the tsunami of the Chilean earthquake, 1960: Technical Report of the Japan Meteorological Agency No. 8, 389 p. [in Japanese; cited on p. 46 as JMA, 1961]. {46, 51, 55, 56, 83, 85, 89}

Japan Meteorological Agency, 1996, Tide tables for the year 1997: Japan Meteorological Agency, 271 p. [in Japanese and English]. {48}

Jishin Yochi Sōgō Kenkyū Shinkōkai Jishin Chōsa Kenkyū Sentā [Association for the Development of Earthquake Prediction, Earthquake Research Center], undated, Daijishin no ato, yoshin wa dō naru ka [What aftershocks will follow a big earthquake?]: prepared for Kagaku Gijutsu Chō [Science and Technology Agency] [in Japanese]. {41}

Johnson, A.C., 1990, An earthquake strength scale for the media and the public: Earthquakes and Volcanoes, v. 22, p. 214-216. {98}

Johnson, J.M., and Satake, K., 1999, Asperity distribution of the 1952 great Kamchatka earthquake and its relation to future earthquake potential in Kamchatka: Pure and Applied Geophysics, v. 154, p. 541-553. {49}

Johnson, J.M., Tanioka, Y., Ruff, L.J., Satake, K., Kanamori, H., and Sykes, L.R., 1994, The 1957 great Aleutian earthquake: Pure and Applied Geophysics, v. 142, p. 3-28. {98}

Johnson, S.Y., Dadisman, S.V., Mosher, D.C., Blakely, R.J., and Childs, J.R., 2001, Active tectonics of the Devils Mountain fault and related structures, northern Puget lowland and eastern Strait of Juan de Fuca region, Pacific Northwest: U.S. Geological Survey Professional Paper 1643, 45 p. {104}

K

Kaizuka, S., Koike, K., Endō, K., Yamazaki, H., and Suzuki, T., editors, 2000, Kanto, Izu, Ogasawara [v. 4 of Nihon no chikei (Regional geomorphology of the Japanese Islands)]: Tokyo, University of Tokyo Press, 349 p. [in Japanese]. {66}

Kajiura, K., Hatori, T., Aida, I., and Koyama, M., 1968, A survey of a tsunami accompanying the Tokachi-oki earthquake of May, 1968: Bulletin of the Earthquake Research Institute, v. 46, p. 1397-1413 [in Japanese with English abstract and figures]. {37, 51, 59}

Kanamori, H., 1977, The energy release in great earthquakes: Journal of Geophysical Research, v. 82, p. 2981-2987. {98}

Kanamori, H., and Heaton, T.H., 1996, The wake of a legendary earthquake: Nature, v. 379, p. 203-204 [commentary introducing report by Satake and others (1996)]. {94}

Kanamori, H., and McNally, K.C., 1982, Variable rupture mode of the subduction zone along the Ecuador-Colombia coast: Bulletin of the Seismological Society of America, v. 72, p. 1241-1253. {101}

Kato, S., 1979a, A history of Japanese literature; 2, the years of isolation: Tokyo, Kodansha, 230 p. [Sanderson, D., translator] {63}

Kato, T., 1979b, Crustal movements in the Tohoku district, Japan, during the period 1900-1975, and their tectonic implications: Tectonophysics, v. 60, p. 141-167. {65}

Katō, Y., Suzuki, Z., Nakamura, K., Takagi, A., Emura, K., Ito, M., and Ishida, H., 1961, The Chile tsunami of May 24, 1960 observed along the Sanriku coast, Japan, in Report on the Chilean tsunami of May 24, 1960, as observed along the coast of Japan: Tokyo, Maruzen Co., Ltd., p. 67-76. {51}

Kawana, N., 1984, Kinsei Nihon suiunshi no kenkyū [Studies of water-transport history of early modern Japan]: Tokyo, Yūzan Kaku, 428 p. [in Japanese]. {66}

Keightley, D.N., 1978, Sources of Shang history; the oracle-bone inscriptions of Bronze Age China: Berkeley, University of California Press, 281 p. {100}

Kelsey, H.M., Witter, R.C., and Hemphill-Haley, E., 2002, Plate-boundary earthquakes and tsunamis of the past 5500 years, Sixes River estuary, southern Oregon: Geological Society of America Bulletin, v. 114, p. 298-314. {16, 22, 101}

Kelsey, H.M., Nelson, A.R., Hemphill-Haley, E., and Witter, R.C., 2005, Tsunami history of an Oregon coastal lake reveals a 4600 yr record of great earthquakes on the Cascadia subduction zone: Geological Society of America Bulletin, v. 117, p. 1009-1032, doi: 10.1130/B25452.1. {101}

Kerr, R.A., 1995, Faraway tsunami hints at a really big Northwest quake: Science, v. 267, p. 962 [news story on research later published by Satake and others (1996)]. {94}

Kin'no, S., editor, 1981, Ezu ni miru hansei jidai no Kesen [Picture maps of Kesen in the time of Sendai-han]: Morioka, Kumagai Insatsu Shuppan-bu, 97 p. [in Japanese]. {81}

Kitamura, N., Kotaka, T., and Kataoka, J., 1961a, Ōfunato-Shizugawa chiku chōsahan [Ōfunato-Shizugawa area investigation group], in Report on the Chilean tsunami of May 24, 1960, as observed along the coast of Japan: Tokyo, Maruzen Co., Ltd., p. 234-244 [in Japanese]. {51, 56}

Kitamura, N., Kotaka, T., and Kataoka, J., 1961b, Ōfunato - Shizugawa chiku [Region between Ōfunato and Shizugawa], in Kon'no, E., editor, Geological observations of the Sanriku coastal region damaged by tsunami due to the Chile earthquake in 1960: Contributions to the Institute of Geology and Paleontology of Tohoku University, v. 52, p. 28-40 [in Japanese with English abstract, figures, and tables]. {19}

Kon'no, E., editor, 1961, Geological observations of the Sanriku coastal region damaged by tsunami due to the Chile earthquake in 1960: Contributions to the Institute of Geology and Paleontology of Tohoku University, v. 52, p. 1-40 [in Japanese with English abstract, figures, and tables]. {56}

Kroeber, A.L., 1976, Yurok myths: Berkeley, University of California Press, 488 p. {20}

L

Lander, J.F., Lockridge, P.A., and Kozuch, M.J., 1993, Tsunamis affecting the west coast of the United States, 1806-1992: National Oceanic and Atmospheric Administration, National Geophysical Data Center Key to Geophysical Records Documentation no. 29, 242 p. {11, 43, 49, 80, 91}

Lay, T., Kanamori, H., Ammon, C.J., Nettles, M., Ward, S.N., Aster, R.C., Beck, S.L., Bilek, S.L., Brudzinski, M.R., Butler, R., DeShon, H.R., Ekström, G., Satake, K., and Sipkin, S., 2005, The great Sumatra-Andaman earthquake of 26 December 2004: Science, v. 308, p. 1127-1133. {5, 98}

Leonard, L.J., Hyndman, R.D., and Mazzotti, S., 2004, Coseismic subsidence in the 1700 great Cascadia earthquake; coastal estimates versus elastic dislocation models: Geological Society of America Bulletin, v. 116, p. 655-670; doi: 10.1130/B25369.1. {16}

Liu, H., and Qiao, T., 1984, Liquefaction potential of saturated sand deposits underlying foundation of structure, in Proceedings of the Eighth World Conference on Earthquake Engineering, v. 3, p. 199-206. {23}

Lockridge, P.A., 1985, Tsunamis in Peru-Chile: National Oceanic and Atmospheric Administration, World Data Center A for Solid Earth Geophysics, Report SE 39, 97 p. {54, 94}

Lockridge, P.A., Whiteside, L.S., and Lander, J.F., 2002, Tsunamis and tsunami-like waves of the eastern United States: Science of Tsunami Hazards, v. 20, p. 120-157 [http://epubs.lanl.gov/tsunami/5103.pdf]. {11}

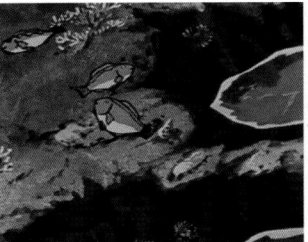

The sea withdraws. The villagers remain on low ground, too far to hear their headman. How can he save them?

Lomnitz, C., 1970, Major earthquakes and tsunamis in Chile during the period 1535 to 1955: Geologische Rundschau, v. 59, p. 938-960. {94}

Lowe, D.R., 1975, Water escape structures in coarse-grained sediment: Sedimentology, v. 22, p. 157-204. {23}

Ludwin, R.S., Dennis, R., Carver, D., McMillan, A.D., Losey, R., Clague, J.[J.], Jonientz-Trisler, C., Bowechop, J., Wray, J., and James, K., 2005, Dating the 1700 Cascadia earthquake; great coastal earthquakes in native stories: Seismological Research Letters, v. 76, p. 140-148. {12}

M

Maritime Safety Agency, 1998, Heisei 11 nen, chō seki hyō (1999 Tide Tables), Maritime Safety Agency Publication no. 781, v. 1. {73, 83, 88}

Mazzotti, S., and Adams, J., 2004, Variability of near-term probability for the next great earthquake on the Cascadia subduction zone: Bulletin of the Seismological Society of America, v. 94, p. 1954-1959. {101}

McCaffrey, R., and Goldfinger, C., 1995, Forearc deformation and great subduction earthquakes: implications for Cascadia offshore earthquake potential: Science, v. 267, p. 856-859. {25}

McCulloch, D.S., and Bonilla, M.G., 1970, Effects of the earthquake of March 27, 1964, on the Alaska Railroad: U.S. Geological Survey Professional Paper 545-D, 161 p. {14}

McDonald, L., 1972, Swan among the Indians; life of James G. Swan, 1818-1900: Portland, Oregon, Binfords and Mort, 233 p. {12}

Melbourne, T.I., Szeliga, W.M., Miller, M.M., and Santillan, V.M., 2005, Extent and duration of the 2003 Cascadia slow earthquake: Geophysical Research Letters, v. 32, L04301, doi:10.1029/2004GL021790. {99}

Miller, M.M., Melbourne, T., Johnson, D.J., and Sumner, W.Q., 2002, Periodic slow earthquakes from the Cascadia subduction zone: Science, v. 295, p. 2423. {99}

Minor, R., and Grant, W.C., 1996, Earthquake-induced subsidence and burial of late Holocene archaeological sites, northern Oregon coast: American Antiquities, v. 61, p. 772-781. {20}

Miyako-shi Kyōiku I'inkai [Miyako City Board of Education], editor, 1981, Miyako-shishi, gyogyō kōeki [Miyako city history, fisheries]: Miyako-shi [Miyako city], 550 p. [in Japanese]. {71}

Miyako-shi Kyōiku I'inkai [Miyako City Board of Education], editor, 1991, Miyako-shishi, nenpyo [Miyako city history, chronology]: Miyako-shi [Miyako city], 615 p. [in Japanese]. {38}

Mofjeld, H.O., Foreman, M.G.G., and Ruffman, A., 1997, West coast tides during Cascadia subduction zone tsunamis: Geophysical Research Letters, v. 24, p. 2215-2218. {83}

Mombushō Shinsai Yobō Hyōgikai [Ministry of Education, Earthquake Disaster Prevention Committee], editor, 1943, Zōtei Dai Nihon jishin shiryō, dai 2 kan [Additional materials on historical earthquakes in Imperial Japan, volume 2] [reprinted in 1975, Tokyo, Meihōsha], 754 p. [in Japanese]. {40, 62, 112}

Mori, K., editor, 1963, Iwate kenshi [History of Iwate prefecture], v. 5, Iwate-kenshi kinsei hen 2 [Early modern history of Iwate prefecture], Morioka-han: Morioka-shi, Tōryō Insatsu, 1,590 p. [in Japanese]. {45, 61}

Mori, K., 1972, Iwate-ken no rekishi [History of Iwate prefecture]: Tokyo, Yamakawa Shuppansha (series: Kenshi shiriizu 3), 320 p. [in Japanese]. {44, 61}

Mori, K., 1983, Kunohe chihō-shi [Local history of Kunohe], Nihon hekichi no shiteki kenkyū [Historical studies of rural Japan], Mori Kahe'e chosakushū dai-9-kan [Collected works of Mori Kahe'e, book 9]: Tokyo, Hōsei Daigaku Shuppan-kyoku [Hōsei University Press], 1,351 p. (reprinted from Kunohe chihō shi, published in 1970 by Kunohe Chihōshi Kankōkai [Kunohe Local-History Publishing Association] of Kuji, Iwate Prefecture) [in Japanese]. {51, 52}

Morioka-shi Chūō Kōminkan [Morioka City Central Community Center], editor, 1998, Nambu Morioka han no ōezu, yomigaeru Edo jidai no fūkei [Large drawings of Nambu- or Morioka-han: Edo-period landscapes brought back to life]: Morioka-shi Chūō Kōminkan (Morioka City Central Community Center), 36 p. [in Japanese; booklet about the drawings]. {44, 45}

Morioka-shi Kyōiku I'inkai [Morioka City Board of Education] and Morioka-shi Chūō Kōminkan [Morioka City Community Center], editors, 1986-2001, Morioka-han zassho: Morioka, Kumagai Insatsu Shuppan-bu, 15 volumes [in Japanese; volume 7, 1083 p., published 1993, entries span Genroku 11-15]. {44}

Morris, I., translator and editor, 1971, The pillow book of Sei Shōnagon: Baltimore, Harmondsworth, Penguin, 1971, 411 p. [reprinted in 1991, by Columbia University Press, New York] {43}

Moss, M.L., and Erlandson, J.M., 1998, Comparative chronology of Northwest Coast fishing features, in Bernick, K., editor, Hidden dimensions; the cultural significance of wetland archaeology: Vancouver, British Columbia, UBC Press, p. 180-198. {21}

Murty, T.S., 1977, Seismic sea waves; tsunamis: Ottawa, Canada, Department of Fisheries and the Environment, Fisheries and Marine Service, 337 p. {11}

Myers, E.P., III, Baptista, A.M., and Priest, G.R., 1999, Finite element modeling of potential Cascadia subduction zone tsunamis: Science of Tsunami Hazards, v. 17, p. 3-18. [http://epubs.lanl.gov/5071.pdf]. {103}

N

Nagaoka, T., 1986, Iwate-ken no kyōiku-shi [History of education in Iwate prefecture]: Kyoto, Shibunkaku Shuppan, 366 p. [in Japanese]. {45}

Nagoya-shi Kyōiku I'inkai [Nagoya City Board of Education], 1965-1969, Ōmu rōchu ki: Nagoya, Nagoya-shi Kyōiku I'inkai, 4 volumes [in Japanese]. {72}

Naitō, A., and Hozumi, K., 1982, Edo no machi (ge) kyodai toshi no hatten [Edo, the growth of a giant city (second of two volumes)]: Tokyo, Sōshisha, 96 p. [in Japanese]. {61}

Naito [Naitō], A., and Hozumi, K., 2003, Edo, the city that became Tokyo: Tokyo, Kodansha, 211 p. [an English combination of the two volumes from 1982; Horton, H.M., translator]. {61, 72, 76}

Nakaminato Shishi Hensan I'inkai [Nakaminato City-History Compilation Committee], 1993, Nakaminato-shi shiryō dai 14 shū (kinsei hasen hen) [Materials about the history of Nakaminato city, v. 14 (early modern shipwrecks)]: Nakaminato-shi, 366 p. [in Japanese]. {66, 68, 71}

Nakamura, K., and Watanabe, H., 1961, Tsunami forerunner observed in case of the Chile tsunami of 1960, in Report on the Chilean tsunami of May 24, 1960, as observed along the coast of Japan: Tokyo, Maruzen Co., Ltd., p. 82-99. {46}

Nelson, A.N., and Haig, J.H., 1997, The new Nelson; Japanese-English character dictionary, revised edition: Rutland, Vermont, and Tokyo, Charles E. Tuttle, 1,600 p. {v, 29, 43}

Nelson, A.R., Atwater, B.F., Bobrowsky, P.T., Bradley, L.-A., Clague, J.J., Carver, G.A., Darienzo, M.E., Grant, W.C., Krueger, H.W., Sparks, R., Stafford, T.W., and Stuiver, M., 1995, Radiocarbon evidence for extensive plate-boundary rupture about 300 years ago at the Cascadia subduction zone: Nature, v. 378, p. 371-374. {21, 25}

Nelson, A.R., Johnson, S.Y., Kelsey, H.M., Wells, R.E., Sherrod, B.L., Pezzopane, S.K., Bradley, L.-A., Koehler, R.D., III, and Bucknam, R.C., 2003, Late Holocene earthquakes on the Toe Jam Hill Fault, Seattle fault zone, Bainbridge Island, Washington: Geological Society of America Bulletin, v. 115, p. 1388-1403, doi: 10.1130/B25262.1. {104}

Nelson, A.R., Asquith, A.C., and Grant, W.C., 2004, Great earthquakes and tsunamis of the past 2000 years at the Salmon River estuary, central Oregon coast, USA: Bulletin of the Seismological Society of America, v. 94, p. 1276-1292. {16}

Nihon Koten Bungaku Daijiten Henshū I'inkai, editor, 1983, Nihon koten bungaku daijiten, jyōkan [Encyclopedia of Japanese classical literature, v. 1]: Tokyo, Iwanami Shoten, 694 p. {29}

Ninomiya, S., 1960, Tsunami in Tōhoku coast induced by earthquake in Chile; a chronological review: Tohoku Kenkyu [Tōhoku Kenkyū; Tōhoku Research], v. 10, no. 6, p. 19-23 [in Japanese with English summary]. {54, 59}

Nishiyama, M., 1997, Edo culture: daily life and diversions in urban Japan, 1600-1868, translated and edited by Gerald Groemer: Honolulu, University of Hawai'i Press, 309 p. {63}

O

Obermeier, S.F., and Dickenson, S.E., 2000, Liquefaction evidence for the strength of ground motions resulting from late Holocene Cascadia subduction earthquakes, with emphasis on the event of 1700 A.D.: Bulletin of the Seismological Society of America, v. 90, p. 876-896. {22, 23}

Ōfunato Shiritsu Hakubutsukan [Ōfunato City Museum], editor and publisher, 1990, Sanriku-engan jishin tsunami nenpyo — Tōhoku-chiho taiheiyo-gawa ni okeru rekishi-jishin, rekishi-tsunami [Chronology of earthquakes and tsunamis on the Sanriku coast — historical earthquakes and tsunamis on Pacific Ocean shores of Tōhoku], 132 p. [in Japanese]. {44, 54}

Ōfunato Shishi Henshū I'inkai [Ōfunato City History Publication Committee], 1978, Ōfunato-shishi dai 3 kan 1 shiryō hen [History of Ōfunato city, volume 3, part 1]: Ōfunato city, 631 p. [in Japanese]. {81}

Omote, S., and Komaki, S., 1961, Ōtsuchi, Yoshihama kan [Between Ōtsuchi and Yoshihama], in Report on the Chilean tsunami of May 24, 1960, as observed along the coast of Japan: Tokyo, Maruzen Co., Ltd., 1961, p. 263-272 [in Japanese]. {65}

Onuki, Y., Shibata, T., and Mii, H., 1961, Tarō - Kamaishi chiku [Region between Tarō and Kamaishi] in Kon'no, E., editor, Geological observations of the Sanriku coastal region damaged by tsunami due to the Chile earthquake in 1960: Contributions to the Institute of Geology and Paleontology of Tohoku University, v. 52, p. 3-27 [in Japanese with English abstract, figures, and tables]. {19}

Oppenheimer, D.H., Beroza, G.C., Carver, G.A., Dengler, L., Eaton, J.P., Gee, L., Gonzalez, F.I., Jayko, A.S., Li, W.H., Lisowski, M., Magee, M.E., Marshall, G.A., Murray, M.H., McPherson, R., Romanowicz, B., Satake, K., Simpson, R.W., Somerville, P.G., Stein, R.S., and Valentine, D., 1993, The Cape Mendocino, California, earthquakes of April 1992—subduction at the triple junction: Science, v. 261, p. 433-438. {18}

Oregon Sea Grant and Oregon State Marine Board, 1999, Boating in Oregon coastal waters: Corvallis and Salem, Oregon Sea Grant and Oregon State Marine Board, 48 p. {73}

Oreskes, N., with Le Grand, H., editors, 2003, Plate tectonics; an insider's history of the modern theory of the Earth: Boulder, Colorado, Westview Press, 424 p. {8}

Ota, Y., and Omura, A., 1991, Late Quaternary shorelines in the Japanese Islands: Daiyonki Kenkyū (The Quaternary Research), v. 30, p. 175-186. {65}

Ōtsuchi-chō Kyōiku I'inkai [Ōtsuchi Town Board of Education], editor, 1961, Chiri jishin tsunami shi [Report on the tsunami from the Chilean earthquake]: Ōtsuchi-chō [Ōtsuchi town], 158 p. [in Japanese]. {65}

Ōuchi, C., editor, 1943, Mito ryōnai de nanpasen [Shipwrecks in the territory of Mito-han]: Mito, Ibaraki-ken Suisan-kai [Ibaraki Prefecture Fishery Association], v. 2 of Ibaraki-ken suisan-shi [History of the fishing industry in Ibaraki prefecture], 553 p. [in Japanese]. {62}

Ovenshine, A.T., Lawson, D.E., and Bartsch-Winkler, S.R., 1976, The Placer River Silt—an intertidal deposit caused by the 1964 Alaska earthquake: Journal of Research of the U.S. Geological Survey, v. 4, p. 151-162. {14}

Owen, G., 1987, Deformation processes in unconsolidated sands, in Jones, M.E., and Preston, R.M.F., editors, Deformation in sediments and sedimentary rocks: Oxford, Blackwell, Geological Society Special Publication 29, p. 11-24. {23}

Ozawa, S., Hashimoto, M., and Tada, T., 1997, Vertical crustal movements in the coastal areas of Japan estimated from tidal observations: Bulletin of the Geographical Survey Institute, v. 43, p. 1-21. {65, 91}

P

Pararas-Carayannis, G., and Calebaugh, J.P., 1977, Catalog of tsunamis in Hawaii: National Oceanic and Atmospheric Administration, World Data Center A for Solid Earth Geophysics, Report SE-4, 78 p. {54}

Parise, F., editor, 1982, The book of calendars: New York, Facts on File, 387 p. [reprinted in 2002 as "The book of calendars; conversion tables for ancient, African, Near Eastern, Indian, Asian, Central American and Western Calendars," by Gorgias Press, Piscataway, New Jersey, 387 p.] {43}

Parsons, T., Tréhu, A.M., Leutgert, J.H., Miller, K., Kilbride, F., Wells, R.E., Fisher, M.A., Flueh, E., ten Brink, U.S., and Christensen, N.I., 1998, A new view into the Cascadia subduction zone and volcanic arc: implications for earthquake hazards along the Washington margin: Geology, v. 26, p. 199-202. {104}

Pascoe, L.C., 1991, Encyclopaedia of dates and events, 3rd edition: London, Hodden and Stoughton, 827 p. {5}

The headman torches his harvested rice. Villagers rush uphill to fight the blaze. Awaiting stragglers, he says, "Let it burn!"

Peters, R., Jaffe, B., Gelfenbaum, G., and Peterson, C., 2003, Cascadia tsunami deposit database: U.S. Geological Survey Open-file Report 03-13, 24 p. [http://geopubs.wr.usgs.gov/open-file/of03-13/]. {18}

Peterson, C.D., 1997, Coseismic paleoliquefaction evidence in the central Cascadia margin, USA: Oregon Geology, v. 59, p. 51-74. {22}

Peterson, M.D., Cramer, C.H., and Frankel, A.D., 2002, Simulations of seismic hazard for the Pacific Northwest of the United States from earthquakes associated with the Cascadia subduction zone: Pure and Applied Geophysics, v. 159, p. 2147-2168. {101, 105}

Plafker, G., 1969, Tectonics of the March 27, 1964 Alaska earthquake: U.S. Geological Survey Professional Paper 543-I, 74 p. {9, 14}

Plafker, G., and Savage, J.C., 1970, Mechanism of the Chilean earthquakes of May 21 and 22, 1960: Geological Society of America Bulletin, v. 81, p. 1001-1030. {11, 19}

Plafker, G., Lajoie, K.R., and Rubin, M., 1992, Determining recurrence intervals of great subduction earthquakes in southern Alaska using radiocarbon dating, in Taylor, R.E., Long., A., and Kra, R.S., editors, Radiocarbon after four decades; an interdisciplinary perspective: New York, Springer-Verlag, p. 436-453. {95}

Portinaro, P., and Knirsch, F., 1987, The cartography of North America: New York, Facts on File, 319 p. {5}

Pratt, T.L., Brocher, T.M., Weaver, C.S., Creager, K.C., Snelson, C.M., Crosson, R.S., Miller, K.C., and Tréhu, A.M., 2003, Amplification of seismic waves by the Seattle basin, Washington State: Bulletin of the Seismological Society of America, v. 93, p. 533-545. {104}

Priest, G.R., Myers, E.[P., III], Baptista, A.[M.], Kamphaus, R., and Peterson, C.D., 1997, Tsunami hazard map of the Yaquina Bay area, Lincoln County, Oregon: Oregon Department of Geology and Mineral Industries Interpretive Map Series IMS-2, scale 1:12,000. {103}

Priest, G.R., Myers, E.[P., III], Baptista, A.[M.], Kamphaus, R.A., Fiedorowicz, B.K., Peterson, C.D., and Horning, T.S., 1998, Tsunami hazard map of the Seaside-Gearhart area, Clatsop County, Oregon: Oregon Department of Geology and Mineral Industries Interpretive Map Series IMS-3, scale 1:12,000. {103}

Priest, G.R., Myers, E.[P., III], Baptista, A.[M.], Erdakos, G., and Kamphaus, R., 1999a, Tsunami hazard map of the Astoria area, Clatsop County, Oregon: Oregon Department of Geology and Mineral Industries Interpretive Map Series IMS-11, scale 1:24,000, 4 p. {103}

Priest, G.R., Myers, E.[P., III], Baptista, A.[M.], and Kamphaus, R., 1999b, Tsunami hazard map of the Warrenton area, Clatsop County, Oregon: Oregon Department of Geology and Mineral Industries Interpretive Map Series IMS-12, 1 sheet, scale 1:24,000, with 5 p. text. {103}

Priest, G.R., Myers, E.P., III, Baptista, A.M., Flueck, P., Wang, K., and Peterson, C.D., 2000a, Source simulation for tsunamis: lessons learned from fault rupture modeling of the Cascadia subduction zone, North America: Science of Tsunami Hazards, v. 18, p. 77-106 [http://epubs.lanl.gov/5082.pdf]. {103}

Priest, G.R., Myers, E.[P., III], Baptista, A.[M.], and Kamphaus, R., 2000b, Tsunami hazard map of the Gold Beach area, Curry County, Oregon: Oregon Department of Geology and Mineral Industries Interpretive Map Series IMS-13, scale 1:12,000, 5 p. {103}

Priest, G.R., Allan, J.C., Meyers, E.P., III, Baptista, A.M., and Kamphaus, R., 2002, Tsunami hazard map of the Coos Bay area, Coos County, Oregon: Oregon Department of Geology and Mineral Industries Interpretive Map Series IMS-21 [CD-ROM]. {103}

R

Raff, A.D., and Mason, R.G., 1961, Magnetic survey off the west coast of North America, 40°N to 52°N latitude: Geological Society of America Bulletin, v. 72, p. 1267-1270. {8}

Riddihough, R., 1984, Recent movements of the Juan de Fuca plate system: Journal of Geophysical Research, v. 89, p. 6980-6994. {8}

Rogers, G.C., 1988, An assessment of the megathrust earthquake potential of the Cascadia subduction zone: Canadian Journal of Earth Sciences, v. 25, p. 844-852. {8}

Rogers, G.[C.], and Dragert, H., 2003, Episodic tremor and slip on the Cascadia subduction zone; the chatter of silent slip: Science, v. 300, p. 1942-1944, doi:10.1126/science.1084783. {99}

Rogers, A.M., Walsh, T.J., Kockelman, W.J., and Priest, G.R., 1996, Earthquake hazards in the Pacific Northwest—an overview, in Rogers, A.M., Walsh, T.J., Kockelman, W.J., and Priest, G.R., editors, Assessing earthquake hazards and reducing risk in the Pacific Northwest, volume 1: U.S. Geological Survey Professional Paper 1560, p. 1-54. {8}

S

Sakudō, Y., 1990, The management practices of family business, in Nakane, C., and Ōishi, S., editors, Tokugawa Japan, the social and economic antecedents of modern Japan [W.B. Hauser, translator, Totman, C., editor of translation]: University of Tokyo Press, p. 147-166. {45}

Satake, K., 2002, Edo-period seismicity along the Kuril Trench estimated from historic documents in Tohoku and Kanto regions: Rekishi Jishin [Historical Earthquakes], v. 18, p. 18-33 [in Japanese with English title and abstract]. {44}

Satake, K. Okada, M., and Abe, K., 1988, Tide gauge response to tsunamis; measurements at 40 tide gauge stations in Japan: Journal of Marine Research, v. 46, p. 557-571. {46}

Satake, K., Shimazaki, K., Tsuji, Y., and Ueda, K., 1996, Time and size of a giant earthquake in Cascadia inferred from Japanese tsunami record of January 1700: Nature, v. 379, p. 246-249. {43, 80, 94}

Satake, K., Wang, K., and Atwater, B.F., 2003, Fault slip and seismic moment of the 1700 Cascadia earthquake inferred from Japanese tsunami descriptions: Journal of Geophysical Research, v. 108, 2325, doi: 10.1019/2003JB002521. {37, 43, 48, 75, 98}

Satō, T., 1988, Nakamura-han, in Kimura, M., Fujino, T., and Murakami, T., editors, Han-shi daijiten dai 1-kan, Hokkaido, Tōhoku hen [Encyclopedia of the history of ruling clans and their Edo-period domains, v. 1, Hokkaido and Tōhoku]: Tokyo, Yūzan Kaku, p. 157-169 [in Japanese]. {69}

Savage, J.C., and Thatcher, W., 1992, Interseismic deformation at the Nankai trough, Japan, subduction zone: Journal of Geophysical Research, v. 97, p. 11,117-11,135. {91}

Sawai, Y., Satake, K., Kamataki, T., Nasu, H., Shishikura, M., Atwater, B.F., Horton, B.P., Kelsey, H.M., Nagumo, T., and Yamaguchi, M., 2004, Transient uplift after a 17th-century earthquake along the Kuril subduction zone: Science, v. 306, p. 1918-1920, doi:10.1126/science.306.5703.1857c. {65}

Scawthorn, C., and Celebi, M., 1987, Performance characteristics of structures, 1985 Mexico City earthquake, in Cassara, M.A., and Martinez Romero, E., editors, The Mexico earthquakes—1985; factors involved and lessons learned: American Society of Civil Engineers, Proceedings of the international conference sponsored by the Mexican section, ASCE, September 19-21, 1986, p. 216-232. {104}

Schweingruber, F.H., 1988, Tree rings; basics and applications of dendrochronology: Dordrecht, D. Reidel, 276 p. {97}

Seeley, C., 2000, A history of writing in Japan: Honolulu, University of Hawai'i Press, 243 p. {40, 63, 100}

Shennan, I., Long, A.J., Rutherford, M.M., Green, F.M., Innes, J.B., Lloyd, J.M., Zong, Y., and Walker, K.J., 1996, Tidal marsh stratigraphy, sea-level change and large earthquakes; a 5000 year record in Washington, U.S.A.: Quaternary Science Reviews, v. 15, p. 1-37. {100}

Sherrod, B.L, Brocher, T.M., Weaver, C.S., Bucknam, R.C., Blakely, R.J., Kelsey, H.M., Nelson, A.R., and Haugerud, R., 2004, Holocene fault scarps near Tacoma, Washington, USA: Geology, v. 32, p. 9-12, doi:10.1130/G19914.1. {104}

Shimizu, I., 1996, Bōsai kyōiku to "Inamura no hi" [Teaching emergency preparedness and "The rice-sheaf fire"]: Rekishi Jishin [Historical Earthquakes], v. 12, p. 215-221 [in Japanese]. {47}

Shimizu, I., 2003, Jidai o [wo] koete hikari kagayaku Hiro-mura tsunami zu "Inamura no hi" [The "Hiro-mura tsunami" painting of "Inamura no hi" shines through time]: Yobō Jihō [Protection News, published by the General Insurance Association of Japan], v. 215, p. 2 [in Japanese] [http://www.sonpo.or.jp/publish/yobojiho/yj215_02.pdf]. {47}

Shively, D.H., 1991, Popular culture, in Hall, J.W., editor, and McClain, J.L., assistant editor, The Cambridge history of Japan (Hall, J.W., Jansen, M.B., Kanai, M., and Twitchett, D., general editors), volume 4, early modern Japan: Cambridge, U.K., Cambridge University Press, p. 706-770. {63}

Sievers C., H.A., Villegas C., G., and Barros, G., 1963, The seismic sea wave of 22 May 1960 along the Chilean coast: Bulletin of the Seismological Society of America, v. 53, p. 1125-1190. {19}

Steel, D., 2000, Marking time; the epic quest to invent the perfect calendar: New York, J. Wiley, 422 p. {43}

Stokes, M.A., and Smiley, T.L., 1968, An introduction to tree-ring dating: Chicago, University of Chicago Press (reprinted in 1996 by University of Arizona Press, Tucson), 73 p. {97}

Stuiver, M., Braziunas, T.F., Becker, B., and Kromer, B., 1991, Climatic, solar, oceanic, and geomagnetic influences on late-glacial and Holocene atmospheric $^{14}C/^{12}C$ change: Quaternary Research, v. 35, p. 1-24. {25}

Stuiver, M., Reimer, P.J., Bard, E., Beck, J.W., Burr, G.S., Hughen, K.A., Kromer, B., McCormac, F.G., v. d. Plicht, J., and Spurk, M., 1998, INTCAL98 Radiocarbon age calibration 24,000 - 0 cal BP: Radiocarbon, v. 40, p. 1041-1083. {25}

Sullivan, W., 1991, Continents in motion; the new Earth debate (2nd edition): New York, American Institute of Physics, 430 p. {8}

Swan, J.G., 1857, The Northwest coast or, three years' residence in Washington Territory: New York, Harper and Brothers, reprinted in 1972 by University of Washington Press, Seattle, 435 p. {12}

Swan, J.G., 1870, The Indians of Cape Flattery, at the entrance to the Strait of Fuca, Washington Territory: Washington, D.C., Smithsonian Contributions to Knowledge v. 16, 108 p. {12}

Swan, J.G., 1971, Almost out of the world; scenes from Washington Territory: Tacoma, Washington State Historical Society, 126 p. [newspaper articles by James G. Swan from the years 1859-1861, selected and edited by W.A. Katz]. {12}

Szeliga, W., Melbourne, T.I., Miller, M.M., and Santillan, V.M., 2004, Southern Cascadia episodic slow earthquakes: Geophysical Research Letters, v. 31, L16602, doi:10.1029/2004GL020824. {99}

T

Takada, K., and Atwater, B.F., 2004, Evidence for liquefaction identified in peeled slices of Holocene deposits along the lower Columbia River, Washington: Bulletin of the Seismological Society of America, v. 94, p. 550-575. {23}

Takahashi, R., and Hatori, T., 1961, A summary report on the Chilean tsunami of 1960, in The Committee for Field Investigation of the Chilean Tsunami of 1960, Report on the Chilean tsunami of May 24, 1960, as observed along the coast of Japan: Tokyo, Maruzen Co., Ltd., p. 23-34. {54}

Takeuchi, R., editor, 1985a, Kadokawa Nihon chimei daijiten [Kadokawa's place names of Japan], v. 3, Iwate-ken [Iwate prefecture]: Tokyo, Kadokawa Shozen, 1282 p. [in Japanese]. {36, 51, 58}

Takeuchi, R., editor, 1985b, Kadokawa Nihon chimei daijiten [Kadokawa's place names of Japan], v. 30, Wakayama-ken [Wakayama prefecture]: Tokyo, Kadokawa Shozen, 1489 p. [in Japanese]. {85}

Tanabe-shi Kyōiku I'inkai [Tanabe City Board of Education], editor, 1987-1991, Kishū Tanabe-machi daichō: Osaka, Seibundō Shuppan, 22 volumes [in Japanese]. {84}

Teramoto, T., Nagata, Y., Sudō, H., and Manabu,T., 1961, Miura-hantō, Hamamatsu kan [Between Miura Peninsula and Hamamatsu], in The Committee for Field Investigation of the Chilean Tsunami of 1960, Report on the Chilean tsunami of May 24, 1960, as observed along the coast of Japan: Tokyo, Maruzen Co., Ltd., p. 321-325 [in Japanese]. {82}

Thatcher, W., 1984, The earthquake deformation cycle at the Nankai Trough, southwest Japan: Journal of Geophysical Research, v. 89, p. 3087-3101. {91}

TEXTBOOK on page 113, "Shōgaku kokugo tokuhon" ["Elementary Japanese-language textbook"], was published by Japan's education ministry, Mombushō, as the fifth volume of a 12-volume set for primary grades. The copy is a reprint from 1971 in the collection of a grade school in Hirogawa, Hiro Shōgakkō. Courtesy of Ikuta Shunji, principal.

VIDEO FRAMES from "Inamura no hi" ["The rice-sheaf fire"], courtesy of Gakken Co., Tokyo.

When the tsunami comes ashore, every villager is standing safely on high ground.

Theberge, A.E., Jr., 2003, 150 years of tides on the western coast: the longest series of tidal observations in the Americas: National Ocean Service, sesquicentennial booklet, 15 p. [http://oceanservice.noaa.gov/topics/navops/ports/150_years_of_tides.pdf]. {91}

Toba, T., and Taka, R., 1961, Matsushima, Nakaminato kan [Between Matsushima and Nakaminato], *in* The Committee for Field Investigation of the Chilean Tsunami of 1960, Report on the Chilean tsunami of May 24, 1960, as observed along the coast of Japan: Tokyo, Maruzen Co., Ltd., p. 303-310 [in Japanese]. {73}

Tokyo Daigaku Jishin Kenkyūsho [University of Tokyo, Earthquake Research Institute], editor and publisher, 1981, Shinshū Nihon jishin shiryō, dai 3 kan bekkan [Newly collected materials on historical earthquakes in Japan, volume 3, appendix], 590 p. [in Japanese]. {86, 89}

Tokyo Daigaku Jishin Kenkyūsho [University of Tokyo, Earthquake Research Institute], editor and publisher, 1993, Shinshū Nihon jishin shiryō, zoku hoi [Newly collected materials on historical earthquakes in Japan, 2nd supplement], 1043 p. [in Japanese]. {51; fingered in photo, opposite}

Toppozada, T.R., Borchardt, G., Haydon, W., Petersen, M., Olson, R., Lagorio, H., and Anvik, T., 1995, Planning scenario in Humboldt and Del Norte Counties, California for a great earthquake on the Cascadia subduction zone: California Division of Mines and Geology Special Publication 115, 157 p. {102}

Totman, C.D., 1967, Politics in the Tokugawa bakufu, 1600-1843: Cambridge, Harvard University Press, 346 p. {61}

Totman, C.[D.], 1989, The green archipelago; forestry in preindustrial Japan: Berkeley, University of California Press, 297 p. {38, 72}

Totman, C.D., 1993, Early modern Japan: Berkeley, University of California Press, 593 p. {29, 45, 63, 72}

Trager, J., 1992, The people's chronology: New York, Henry Holt and Co., 1237 p. {87}

Tremblay, R., 1998, Development of design spectra for long-duration ground motions from Cascadia subduction earthquakes: Canadian Journal of Civil Engineering, v. 25, p. 1078-1090. {104}

Tsuji, Y., 1987, Tsunami-daka to higai no kankei [Relationship between tsunami height and damage]: Rekishi Jishin [Historical Earthquakes], v. 3, p. 239-256 [in Japanese]. {48}

Tsuji, Y., and Ueda, K., 1995, Keichō 16 nen (1611), Enpō 5 nen (1677), Hōreki 12 nen (1763), Kansei 5 nen (1793), oyobi Ansei 3 nen (1856) no kaku Sanriku tsunami no kenshū [Study of the 1611, 1677, 1763, 1793, and 1856 Sanriku tsunamis]: Rekishi Jishin [Historical Earthquakes], v. 11, p. 75-106 [in Japanese]. {37, 51, 59, 64}

Tsuji, Y., Ueda, K., and Satake, K., 1998, Japanese tsunami records from the January 1700 earthquake in the Cascadia subduction zone: Zisin [Journal of the Seismological Society of Japan], v. 51, p. 1-17 [in Japanese with English title, abstract, and captions]. {43, 48, 57, 62, 64, 72, 82, 83, 88, 90}

Tsukahira, T., 1966, Feudal control in Tokugawa Japan: the sankin kōtai system: Cambridge, Mass., Harvard East Asia Monographs 20, 228 p. {61}

Tsumura, K., 1991, "Inamura no hi" to Hiro-mura teibō ["Inamura no hi" and the seawall at Hiro-mura]: Jishin Journal [Earthquake Journal, published by the Association for the Development of Earthquake Prediction], v. 12, p. 22-29 [in Japanese]. {47}

Tsunoda, R., de Bary, W.T., and Keene, D., compilers, 1964, Sources of Japanese tradition, volume 1: New York, Columbia University Press, 506 p. {63}

Tufte, E.R., 1990, Envisioning information: Cheshire, Connecticut, Graphics Press, 126 p. {109}

Tufte, E.R., 2001, The visual display of quantitative information (2nd edition): Cheshire, Connecticut, Graphics Press, 197 p. {109}

U

Uchida, M., 1975, Nihon rekijitsu genten [Handbook of Japanese calendars]: Tokyo, Yūzankaku Shuppan, 560 p. [in Japanese]. {43}

Ueda, K., and Usami, T., 1990, Yushi irai no jishin kaisu no hensen [Changes in the yearly number of historical earthquakes in Japan]: Rekishi Jishin [Historical Earthquakes], v. 6, p. 181-187 [in Japanese]. {63}

UNAM [Universidad Nacional Autónoma de México] Seismology Group, 1986, The September 1985 Michoacan earthquakes; aftershock distribution and history of rupture: Geophysical Research Letters, v. 13, p. 573-576. {9}

Unno, K., 1994, Cartography in Japan, *in* Harley, J.B., and Woodward, D., editors, The history of cartography, volume two, book two, Cartography in the traditional East and East Asian societies: Chicago, University of Chicago Press, p. 346-477. {29, 32}

Unoki, S., and Tsuchiya, M., 1961, Tarō Funakoshi kan [Between Tarō and Funakoshi], in Report on the Chilean tsunami of May 24, 1960, as observed along the coast of Japan: Tokyo, Maruzen Co., Ltd., p. 257-263 [in Japanese]. {37, 49}

Usami, T., 1979a, Study of historical earthquakes in Japan: Bulletin of the Earthquake Research Institute, v. 54, p. 399-439. {62}

Usami, T., 1979b, Contributors to the collection of historical data of Japanese earthquakes—Messrs. Minoru TAYAMA and Kinkichi MUSHA—: Zisin [Journal of the Seismological Society of Japan], v. 32, p. 355-359 [in Japanese]. {62}

Usami, T., 1996, Shimpen Nihon higai jishin sōran, zōho kaitei-ban 416-1995 (Materials for a comprehensive list of destructive earthquakes in Japan, 416-1995, revised and enlarged edition): Tokyo, University of Tokyo Press, 493 p. [in Japanese]. {37, 51, 62, 91}

W

Walker, B.L. 2001, The conquest of Ainu lands; ecology and culture in Japanese expansion, 1590-1800: Berkeley, University of California Press, 332 p. {61}

Walsh, T.J., Combellick, R.A., and Black, G.L., 1995, Liquefaction features from a subduction zone earthquake; preserved examples from the 1964 Alaska earthquake: Washington Division of Geology and Earth Resources, Report of Investigations 32, 80 p. {22}

Walsh, T.J., Caruthers, C.G., Heinitz, A.C., Myers, E.P., III, Baptista, A.M., Erdakos, G.B., and Kamphaus, R.A., 2000, Tsunami hazard map of the southern Washington coast: modeled tsunami inundation from a Cascadia subduction zone earthquake: Washington Division of Geology and Earth Resources, Geologic Map GM-49, scale 1:100,000, with 12-page pamphlet. {103}

Walsh, T.J., Myers, E.P., III, and Baptista, A.M., 2002a, Tsunami inundation map of the Port Angeles, Washington, area: Washington Division of Geology and Earth Resources Open File Report 2002-1, scale 1:24,000.

[http://www.dnr.wa.gov/geology/pdf/ofr02-1.pdf]. {103}

Walsh, T.J., Myers, E.P., III, and Baptista, A.M., 2002b, Tsunami inundation map of the Port Townsend, Washington, area: Washington Division of Geology and Earth Resources Open File Report 2002-2, scale 1:24,000. [http://www.dnr.wa.gov/geology/pdf/ofr02-2.pdf]. {103}

Walsh, T.J., Myers, E.P., III, and Baptista, A.M., 2003a, Tsunami inundation map of the Quileute, Washington, area: Washington Division of Geology and Earth Resources Open File Report 2003-1, 1 sheet, scale 1:24,000 [http://www.dnr.wa.gov/geology/pdf/ofr03-1.pdf]. {103}

Walsh, T.J., Myers, E.P., III, and Baptista, A.M., 2003b, Tsunami inundation map of the Neah Bay, Washington, area: Washington Division of Geology and Earth Resources Open File Report 2003-2, 1 sheet, scale 1:24,000 [http://www.dnr.wa.gov/geology/pdf/ofr03-2.pdf]. {103}

Walsh, T.J., Titov, V.V., Venturato, A.J., Mofjeld, H.O., and González, F.I., 2004, Tsunami hazard map of the Bellingham area—modeled tsunami inundation from a Cascadia subduction zone earthquake: Washington Division of Geology and Earth Resources Open File Report 2004-15, scale 1:50,000 [http://www.dnr.wa.gov/geology/pdf/ofr04-15.pdf]. {103}

Walter, L., 1994, Catalogue, in Walter, L., editor, Japan, a cartographic vision; European printed maps from the early 16th to the 19th century: Munich and New York, Prestel-Verlag, 232 p. {31}

Wang, K., Wells, R.E., Mazzotti, S., Dragert, H., Hyndman, R.D., and Sagiya, T., 2003, A revised 3-D dislocation model of interseismic deformation for the Cascadia subduction zone: Journal of Geophysical Research, v. 108, 2026, doi:10.1029/2001JB001227. {99}

Ward, P.L., Page, R.A., Hodgen, L.D., and Troll, J.A., 1989, The Loma Prieta earthquake of October 17, 1989; a brief geologic view of what caused the Loma Prieta earthquake and implications for future California earthquakes—what happened ... what is expected ... what can be done: U.S. Geological Survey pamphlet, 16 p. {109}

Watanabe, H., 1998, Nihon higai tsunami sōran, dai ni-han (Comprehensive list of destructive tsunamis to hit the Japanese islands, 2nd edition): Tokyo, University of Tokyo Press, 238 p. [in Japanese]. {54, 59, 62, 77, 85, 94}

Weischet, W., 1963, Further observations of geologic and geomorphic changes resulting from the catastrophic earthquake of May 1960, in Chile: Bulletin of the Seismological Society of America, v. 53, p. 1237-1258. {11}

Williams, H.F.L., Hutchinson, I., and Nelson, A.R., 2005, Multiple sources for late Holocene tsunamis at Discovery Bay, Washington State, USA: The Holocene, v. 15, p. 60-73, doi:10.1191/0956683605hl784rp. {18}

Williams, N., 1999, Chronology of world history, v. II, 1492-1775, the expanding world: Santa Barbara, Calif., ABC-CLIO, 765 p. {5}

Witter, R.C., Kelsey, H.M., and Hemphill-Haley, E., 2003, Great Cascadia earthquakes and tsunamis of the past 6700 years, Coquille River estuary, southern coastal Oregon: Geological Society of America Bulletin, v. 115, p. 1289-1306. {16, 101}

Wright, C., and Mella, A., 1963, Modifications to the soil pattern of south-central Chile resulting from seismic and associated phenomena during the period May to August 1960: Bulletin of the Seismological Society of America, v. 53, p. 1367-1402. {19}

Y

Yamaguchi, D.K., Atwater, B.F., Bunker, D.E., Benson, B.E., and Reid, M.S., 1997, Tree-ring dating the 1700 Cascadia earthquake: Nature, v. 389, p. 922-923, editors' correction in v. 390, p. 352. {96, 97}

Yamashita, F., 1997, Tsunami: Tokyo, Ayumi Shuppan, 222 p. [in Japanese]. {37, 41, 51, 59}

Yanuma, T., and Tsuji, Y., 1998, Observation of edge waves trapped on the continental shelf in the vicinity of Makurazaki Harbor, Kyushu, Japan: Journal of Oceanography, v. 54, p. 9-18 [http://www.terrapub.co.jp/journals/JO/pdf/5401/54010009.pdf] {86}

Yeats, R.S., 2004, Living with earthquakes in the Pacific Northwest, a survivor's guide, second edition: Corvallis, Oregon State University Press, 400 p. {104}

Yoshida, Y., and Oikawa, K., 1983-1992, Zusetsu Morioka yonhyaku-nen [Four hundred years of Morioka history]: Morioka, Kyodo Bunka Kenkyūkai, 3 volumes, 1,458 p. [in Japanese]. {44}

Yoshinobu, E., 1961, Shinjō-cho ni okeru Ansei Nankai Chiri jishin ni yoru tsunami no takasa no sokutei [Height measurements of the tsunamis in Shinjō ward caused by the Ansei, Nankai, and Chilean earthquakes]: Tanabe bunkazai [Cultural properties of Tanabe city], v. 5, p. 14-19 [in Japanese]. {85, 89}

Z

Zachariasen, J., Sieh, K., Taylor, F.W., Edwards, R.L., and Hantoro, W.S., 1999, Submergence and uplift associated with the giant 1833 Sumatran subduction earthquake: evidence from coral microatolls: Journal of Geophysical Research, v. 104, p. 895-919. {5}

Accounts of the 1700 tsunami form part of the 21-volume, 16,812-page earthquake anthology, "Shinshū Nihon jishin shiryō" (p. 62). Two of the volumes are cited at the top of the facing page, under Tokyo Daigaku Jishin Kenkyūsho.

First volume, shelved at upper right, was published in 1981.

Fingered volume, published in 1993, contains accounts of the orphan tsunami of 1700.

Index 索引

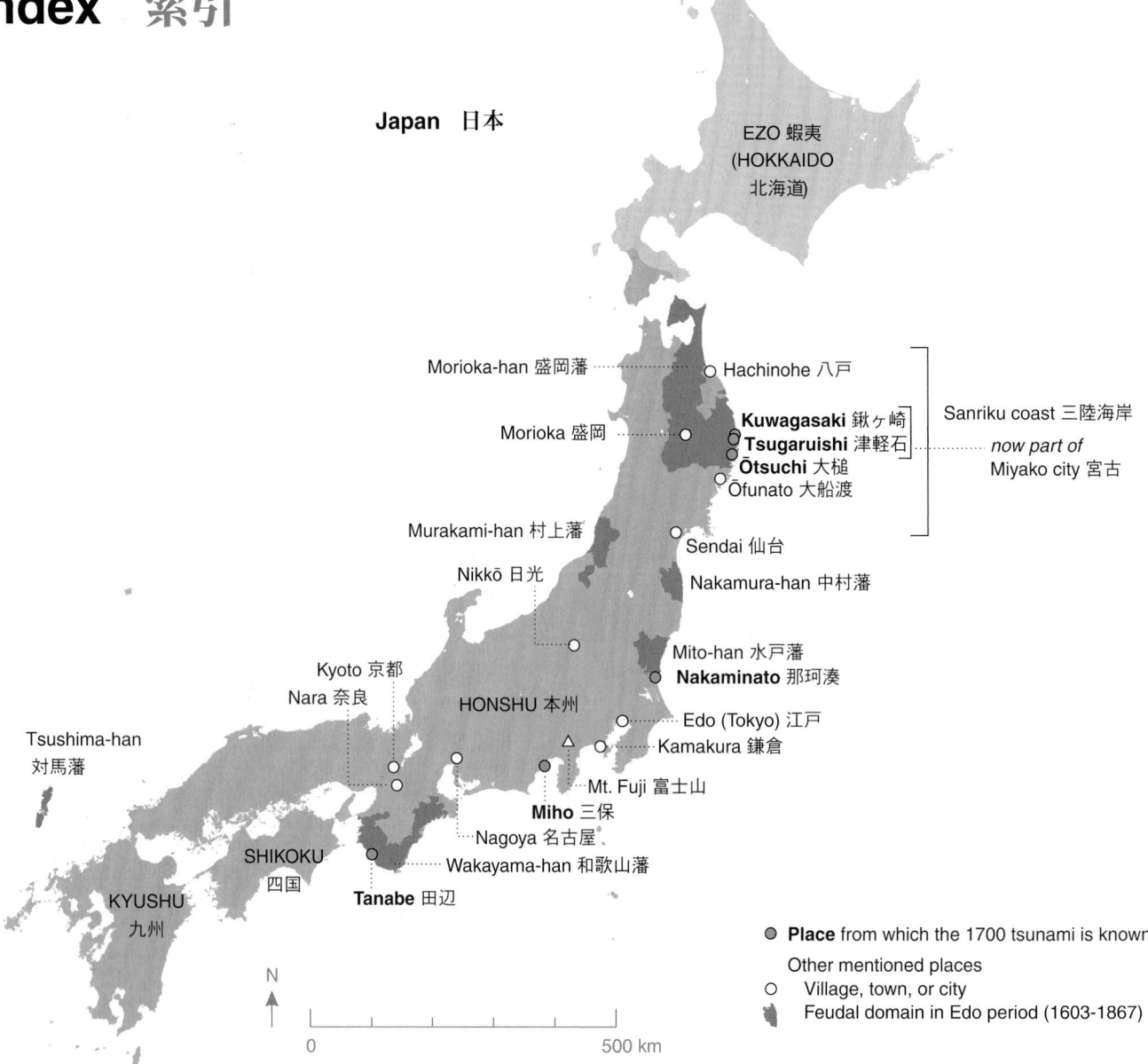

Japan 日本

EZO 蝦夷
(HOKKAIDO
北海道)

Morioka-han 盛岡藩 ········ Hachinohe 八戸

Morioka 盛岡

Kuwagasaki 鍬ヶ崎
Tsugaruishi 津軽石 ········ Sanriku coast 三陸海岸
Ōtsuchi 大槌 *now part of*
Ōfunato 大船渡 Miyako city 宮古

Murakami-han 村上藩 ········ Sendai 仙台

Nikkō 日光 Nakamura-han 中村藩

Mito-han 水戸藩
Kyoto 京都 **Nakaminato** 那珂湊
Nara 奈良
HONSHU 本州 Edo (Tokyo) 江戸
Kamakura 鎌倉
Tsushima-han
対馬藩 Mt. Fuji 富士山
Miho 三保
SHIKOKU Nagoya 名古屋
四国
KYUSHU Wakayama-han 和歌山藩
九州 **Tanabe** 田辺

● **Place** from which the 1700 tsunami is known
Other mentioned places
○ Village, town, or city
Feudal domain in Edo period (1603-1867)

N

0   500 km

A

Aberdeen, town beside Grays Harbor, Washington:
    tsunami-evacuation map, 102n
    tsunami hazards, 102-103
Abiki あびき, term for "unusual seas" in Tanabe account of 1700
      tsunami, 40, 86
Ainu, native people of Ezo (Hokkaido), 61n
Akamae 赤前, village near Tsugaruishi;
    tsunami inundation, 51, 56-57
    sand sheet from 1960 tsunami, 18n
Alaska earthquake of 1964:
    damage by shaking, 9
    enormity, 9, 98

m, on map only; n, in footnote only
Italicized entries are Japanese terms other than place names

    geological records of subsidence, 14-15
    liquefaction features, 22n
    predecessors, 95
    rupture area, 9m
Alaska tsunami of 1964:
    deaths in Oregon and California, 11
    height in Japan, 94-95
    sand in British Columbia, 18n
Archaeological sites, in coastal Washington and Oregon, abandoned
      around 1700, 20-21
Astoria, Oregon, town near mouth of Columbia River; tide-gauge
      record of 1854 Tōkai tsunami, 91
Atonoura 跡ノ浦, village near Tanabe; fields flooded by 1700
      tsunami, 85-86, 90
Ayukawa 鮎川, town on southern Sanriku coast; subsidence recorded
      by tide gauge, 65

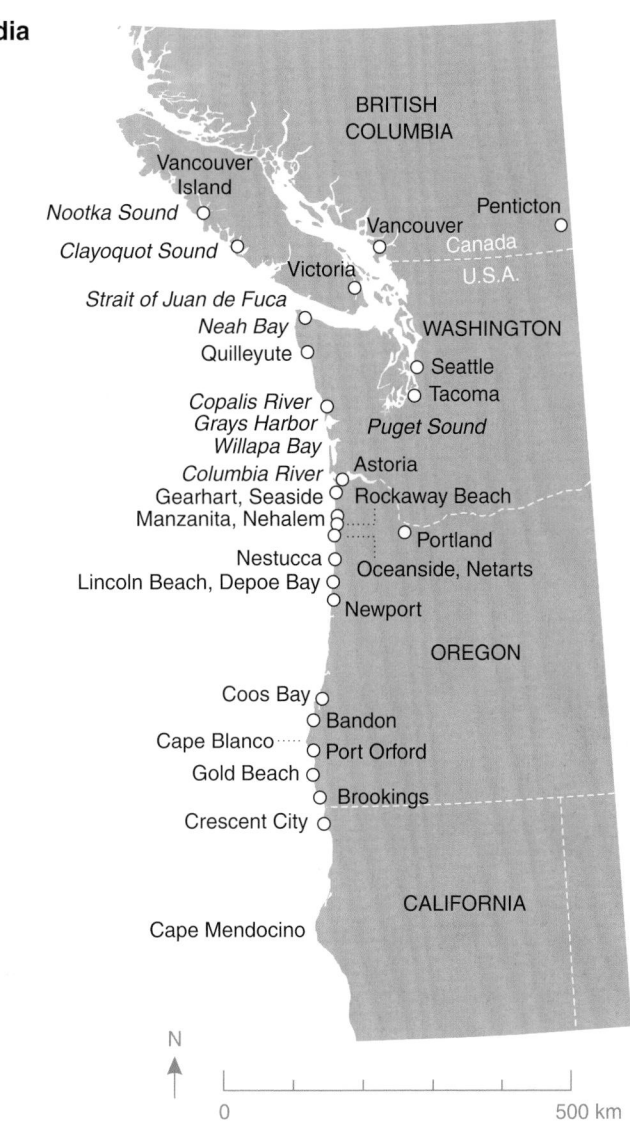

Cascadia

BRITISH COLUMBIA

Vancouver Island

Nootka Sound

Clayoquot Sound

Penticton

Vancouver

Canada

Victoria

U.S.A.

Strait of Juan de Fuca

WASHINGTON

Neah Bay

Quilleyute

Seattle

Tacoma

Copalis River
Grays Harbor
Willapa Bay

Puget Sound

Columbia River

Astoria

Gearhart, Seaside

Rockaway Beach

Manzanita, Nehalem

Portland

Nestucca

Oceanside, Netarts

Lincoln Beach, Depoe Bay

Newport

OREGON

Coos Bay

Bandon

Cape Blanco

Port Orford

Gold Beach

Brookings

Crescent City

CALIFORNIA

Cape Mendocino

N

0   500 km

B

Bache, Alexander Dallas (1806-1867), superintendent, U.S. Coast Survey; report on American tide-gauge records of Japanese tsunamis of December 1854, 91

Bakufu ("tent government"), national military government headed by a shogun; the shogunate:
Fiscal burden to daimyo domains, 61
Shogunal headquarters in Kamakura, Kyoto, and Edo, 63

Balch, Billy, Makah leader (Makah name, Yelakub); told flood tradition, 12-13

Bandon, Oregon, tsunami-evacuation map, 102n

Bay Center, village beside Willapa Bay, Washington; tsunami-evacuation map, 102n

Bellevue, city east of Seattle, Washington; tall buildings and shaking hazards, 104-105

Biak, Indonesia; source region of 1996 tsunami in Japan, 54m, 94

Bookworms, trails in old documents, 32m, 87

Boston, Massachusetts, population in 1700, 5n

British Columbia:
1700 tsunami, geologic evidence, 18m
Clayoquot and Nootka mentioned in Native American flood tradition, 12-13
radiocarbon dating of earthquakes, 101
sand deposited by 1964 Alaska tsunami, 18n
subsidence during earthquakes, geologic evidence, 16m
tall buildings and shaking hazards, 104-105

Brookings, Oregon, tsunami-evacuation map, 102n

C

Calendars, 42 (*see also* Eras, Zodiac)

California:
1700 tsunami, geologic evidence, 18m

* MAJOR JAPANESE EARTHQUAKES are known in Japan by their era name and region; the 1611 event, for instance, is the Keichō Sanriku earthquake. The 1854 events, though traditionally assigned to the Ansei era, predated the start of that era by three weeks.

Date Masamune, daimyo of Sendai-han, is associated with the earliest known writing of 津波 (p. 41).

Courtesy of Sendai Museum

University of Washington Libraries, Special Collections, NA0410

Makah Indians sit for a portrait by J.G. Swan, probably in the 1860s.
In 1864, another Makah told Swan a sea-flood legend the 1700
tsunami may have inspired (p. 12-13).

Ōuchi-family compound in Nakaminato, 1842. The storehouse may have then held Ōuchi-ke "Go-yōdome," the collection of shipwreck documents that refers to the 1700 tsunami (p. 66).

Courtesy of Hitachinaka City

* Like the associated earthquakes (footnote, p. 126), these tsunamis are known by era and region. Complications: Hakuhō is obsolete as an era name, and the 1854 tsunamis predate the Ansei era but are customarily assigned to it.

A wave from Chile in 1960 approaches roofs of Ōfunato, Japan (p. 81).

Courtesy of Ōfunato city